W0262752

Heinz Ulbricht was born in Sommerfeld 1931. He received his Diploma, his Dr. rer. nat. and Dr. sc. nat. at the Rostock University (1956, 1963, 1971) under Prof. Hans Falkenhagen and finished a post graduate study at the Moscow State University. Now he is Professor of Theoretical Physics at the Wilhelm-Pieck-University of Rostock and author (or editor) of two books on continuum mechanics and the theory of selforganization.
Working fields: Thermodynamics and the theory of elektrolytic solutions.

Jürn Schmelzer was born in Grevesmühlen 1951. In 1975 he finished his studies in physics at the Odessa State University, USSR. In 1979 he defended his Dr. rer. nat. under Prof. W. Ebeling at the Rostock University. He finished post graduate studies in Sofia and Oxford. In 1985 he received his Dr. sc. in theoretical physics.
Working fields: Thermodynamics of finite systems, kinetics of phase transitions and thermodynamic description of small clusters.

Reinhard Mahnke was born in Rostock 1952. He received his Diploma 1976 at the Rostock University and his Dr. rer. nat. under Prof. W. Ebeling. He finished a post graduate study at the Latvian State University of Riga (1982/83).
Working fields: Theory of nonlinear dynamical systems especially the kinetic behaviour in self-organizing systems.

Frank Schweitzer was born in Schwaan 1960. He received his Diploma in 1983 at the Rostock University and his Dr. rer. nat. in 1986 after accomplishing a research study directed by Prof. H. Ulbricht and partially by Prof. W. Ebeling in Berlin.
Working fields: Thermodynamics of nucleation in one- and many-component systems and the stochastic description of phase transitions.

Ulbricht, Heinz: Thermodynamics of Finite Systems and the Kinetics of First-Order Phase Transitions/ H. Ulbricht; J. Schmelzer; R. Mahnke; F. Schweitzer. - Leipzig: BSB Teubner (Teubner-Texte zur Physik; 17) - 1. Aufl. - 1988. - 208 S.
NE: Schmelzer, Jürn; Mahnke, Reinhard; Schweitzer, Frank:; GT

ISBN 978-3-322-96428-1   ISBN 978-3-322-96427-4 (eBook)
DOI 10.1007/978-3-322-96427-4

ISSN 0233-0911

1. Auflage
VLN 394-375/89/88 · LSV 1115
Lektor: Dr. rer. nat. Ingrid Lischke

Gesamtherstellung: (52) Nationales Druckhaus, Berlin, Betrieb der VOB National, 1055 Berlin
Bestell-Nr. 666 471 8
02600

TEUBNER–TEXTE zur Physik · Band 17

Herausgeber/Editors: Werner Ebeling, Berlin

Wolfgang Meiling, Dresden

Armin Uhlmann, Leipzig

Bernd Wilhelmi, Jena

Heinz Ulbricht / Jürn Schmelzer / Reinhard Mahnke
Frank Schweitzer

Thermodynamics of Finite Systems and the Kinetics of First-Order Phase Transitions

Phase transitions between different states of matter occur in many
equilibrium and nonequilibrium systems. The kinetics of these phase
changes are studied mainly in the case of condensation of a single
vapour in a supersaturated state by homogeneous nucleation. Based
on thermodynamic investigations of heterogeneous systems the stochastic and deterministic theory of nucleation and growth of the new
phase is derived. The emphasis lies on finite-size-effects which
lead to a depletion of the vapour. The results are mainly explained
in terms of clusters (droplet model) and enlarged to the influence
of external fields. Numerous computer simulations are presented.

1

Phasenübergänge zwischen verschiedenen Aggregatzuständen der Materie
sind in vielen Gleichgewichts- und Nichtgleichgewichtssystemen zu be-
obachten. Im Falle der Kondensation aus einem übersättigten Dampf kann
die Kinetik durch das Modell der homogenen Keimbildung erfaßt werden.
Auf der Grundlage thermodynamischer Untersuchungen heterogener Systeme
wird eine entsprechende stochastische und deterministische Theorie der
Nukleation entwickelt. Dabei führt die Endlichkeit des Systems zu
einer Aufzehrung des Dampfes. Die Ergebnisse werden im Tröpfchenmodell
(Clusterbildung) beschrieben und auf die Berücksichtigung äußerer Fel-
der ausgedehnt. Zahlreiche Computersimulationen verdeutlichen die Re-
sultate.

Dans beaucoup de systémes équilibres et non-équilibres il est á obser-
ver des phases transitoires entre des états d'agrégation différents
de la matiére. Au cas de la condensation d'une vapeur de la situation
sursaturée on peut comprendre la cinétique au moyen du modéle de la
formation de germes homogénes. Sur la base de recherches thermodynami-
ques des systémes hétérogénes a été développé une théorie stochasti-
que et déterministe correspondante au processus de la nucléation. A
la fois la limitation du systéme fait naître une consommation de la
vapeur. Les résultats sont déscrits au moyen du modéle de goutte
(termes of clusters) et de plus on a pris en considération des champs
extérieurs. Des simulations d' ordinateurs rendent clair les résultats.

Transiciones de fase ocurren entre diferentes estados de la materian
tanto en sistemas en el equilibrio como en no-equilibrio. Está estu-
diada la cinética de estas transiciones de fase principalmente en el
caso de la condensación de un vapor supersaturado por medio de la
nucleación. Basándose en investigaciones termodinámicas de sistemas
inhomogéneos se desarrolla la teoria estocástica y determinista del
proceso de nucleación. Se hace énfasis en efectos de superficie que
llevan a la desapariçión del vapor. Los resultados están interpreta-
dos principalmente con términos de cúmulos (droplet model) y genera-
lizados pora el case de un campo externo. Numerosas simulaciones nu-
mérios aclaran los resultados.

Фазовые переходы между различными агрегатными состояниями материи
наблюдаются во многих равновесных и неравновесных системах. В слу-
чае конденсации перенасыщеного пара кинетику процесса можно опи-
сать с помощью модели гомогенного образования зародышей. На основе
термодинамических исследований гетерогенных систем развивается сто-
хастическая и детерминистская теория образования зародышей. При
этом конечность системы ведет к потреблению пара. Результаты опи-
сываются в капельной модели / образование кластеров / и расширя-
ются на случай с учетом влияния внешних полей. Компьюторное мо-
делирование поясняет результаты.

Contents

<u>Foreword</u>

This booklet is devoted to the thermodynamic and kinetic description
of first-order phase transitions. In general, the matter of the world
exists in different phases. Normally phase changes take place in thermodynamic equilibrium, which will be considered here.

Typically,the system is rapidly quenched from a one-phase thermal
equilibrium state to a nonequilibrium situation. During the so-called
equilibrium phase transformation process the quenched supersaturated
system evolves from the nonequilibrium state to an equilibrium one
which consists of two coexisting phases. In a series of books on phase
transitions and critical phenomena (DOMB, GREEN, LEBOWITZ, 1972 - 1983)
an immense amount of material to different aspects of this topic is
summarized.

The other type of phase transitions takes place in systems far from
equilibrium. Due to the nonequilibrium boundary conditions and the fluxes from the environment into the system the final state of this so-
called nonequilibrium phase transition is a stable nonequilibrium situation. Such interesting processes (e.g. pattern formation, multista-
bility) do not appear only in physics but also in chemistry, meteorolo-
gy, biology and many areas of engineering. Concerning questions in this
context we recommend the reader to the monographs by HAKEN (1978), and
EBELING, FEISTEL (1982). An overview of the problems of recent interest
in this field is given in the Proceedings of the Third International
Conference on Irreversible Processes and Dissipative Structures, edited
by EBELING and ULBRICHT (1986). Interesting results for nonlinear dissipative systems dealing with selforganization and turbulence (EBELING,
KLIMONTOVICH, 1984), bistability (MALCHOW, SCHIMANSKY-GEIER, 1985),
nonlinear flows (SCHULTZ-PISZACHICH, 1985) have been published in this
series of Teubner-Texte zur Physik.

Here we limit our considerations to equilibrium phase transitions in
finite systems, mainly to the case of homogeneous nucleation in supersaturated vapours with a constant overall particle number. Therefore
the condensing vapour is rapidly depleted. Some of this ideas on finite systems were carried out in collaboration with Dr. sc. W. Vogelsberger (Jena).

Despite the general thermodynamic approach to nucleation investigated
in this book, some problems remain unsolved for an application to special systems. E.g. a complete theory of nucleation in binary systems is
still in request. The kinetic theories presented here base mainly on
the classical droplet model and,therefore, they are refered to low-su-
persaturated systems. Near the spinodal region new droplet models or

field theoretic nucleation theories of the Ginzburg-Landau or Langer
type must be applied, which will not reviewed in this book (see e.g.
GUNTON et al., 1983). Further we do not consider here the problem of
changing thermodynamic constraints during the phase transition, e.g.
by cooling rates. The mentioned problems give only some examples to
extend or improve the presented theories. Some recent investigations by
different authors are joined in Rostocker Physikalische Manuskripte,
No. 10, edited by ULBRICHT and SCHWEITZER (1987).

The authors would like to thank Prof. W. Ebeling, Dr. sc. L.
Schimansky-Geier and Doz. R. Feistel (Berlin) for pleasant cooperation
and for their continuous support in our work. We are grateful for many
helpful discussions with Prof. P. P. Wegener (New Haven), Prof. I.
Gutzow, Dr. Radost Pascova (Sofia) and Prof. J. S. Rowlinson (Oxford).
One of us (J.S.) acknowledges the hospitality of the Physical Chemistry
Laboratory, Oxford, and the Institute of Physical Chemistry, Sofia,
where portions of this work were accomplished.
The authors thank Dipl.-Phys. A. Budde for his scientific support and
his engagement in preparing the manuscript.
Last but not least we want to thank Mrs. Hiltrud Bahlo who typed the
whole manuscript in a perfect way and Mrs. Gerlind Przybilski who drew
our figures.
Finally we mention the smooth cooperation with the editors of this series as well as with Dr. Ingrid Lischke from Teubner-Verlag Leipzig.

Rostock, June 1987   Heinz Ulbricht
                                        Jürn Schmelzer
                                        Reinhard Mahnke
                                        Frank Schweitzer

1. Introduction

1.1. Types and Classification of Phase Transitions

The condensation of gases, melting of solids, freezing of liquids, the crystallization of glasses, the decomposition of multicomponent solid or liquid solutions, the spontaneous magnetization of ferromagnetics and antiferromagnetics, the appearance of a spontaneous polarization in ferroelectrics, superfluidity and superconductivity are examples of thermodynamic phase transitions. In general, phase transitions can be considered as qualitative variations of the structure or the reaction of a physical system with respect to a change of external parameters as pressure, temperature, electric and magnetic fields.

The investigation of phase transitions has a long history. Let us mention that already Fahrenheit could prove the existence of so-called metastable states, homogeneous states of a system, which are stable with respect to small fluctuations but unstable with respect to fluctuations of a finite size.

Experiments concerning supersaturated systems are difficult to prepare as was already pointed out in the 19th century by James Thomson and Robert von Helmholtz. At this early stage the studies of supersaturated systems were mainly restricted to the condensation of vapours, e.g. the condensation of supersaturated water vapour under different circumstandes. Milestones in the understanding of the thermodynamics and kinetics of metastable states are the important experiments by Laval (supersonic steam nozzle, 1889), Wilson (cloud chamber, 1897), Prandtl (schlieren photography) and others (see e.g. WEGENER, MIRABEL, 1987).

Parallel to these observations the phenomena of capillarity were discussed by Laplace, Thomas Young, William Thomson and other scientists. It was clear before 1800 that gravitational forces were inadequate to explain these properties of liquids. Alternative repulsive and attractive forces must be present. Laplace said that the only condition imposed on these forces was that they were "insensible at sensible distances" (ROWLINSON, WIDOM, 1982). In 1806 he demonstrated that the pressure in a drop of radius r is higher than the surrounding pressure by $2\mathfrak{G}/r$, where $\mathfrak{G}$ is the surface tension of the liquid. Since the first steps by Laplace the quantity $\mathfrak{G}$ is still under discussion not only from the thermodynamic point of view (see Chapter 2) also from statistical mechanical considerations of inhomogeneous systems (HENDERSON, 1986).

In 1878 J. W. Gibbs derived an expression for the work needed to form
a critical droplet in the vapour phase and related equilibrium thermodynamics to the stability of phases. The fundamental work by GIBBS
(1875/78) is one of the basic requirements of a theory of phase transitions (GOODRICH, RUSANOV, 1981).
Another aspect of phase transitions is related to the fundamental kinetic theory of metastable states which goes back in the 1930's to
Volmer, Weber, Becker, Döring, Stranski, Kaischew, Zeldovitsch, Frenkel and others. In 1926 Volmer and Weber gave an first expression for
the rate of formation of critical nuclei per unit volume, which was
later improved and completed by a kinetic factor (Zeldovitsch factor).
In this context we quote from the numerous literature only the origi-
nal monographs   by VOLMER (1939) and FRENKEL (1955).

Based on these investigations together with experimental observations
the nucleation theory as a part of the theory of phase transitions of
different kinds was established. A first classification of possible
types of phase transitions was proposed by P. EHRENFEST (1933). According to this theory one has to divide between phase transitions of
different order, mainly between phase transitions of first and second
order in dependence on the properties of the Gibbs free energy at the
transition point. Phase transitions of first order are characterized
by discontinuities of the inner energy U, the entropy S, the volume
of the system V. In thermodynamic phase transitions of second order
these quantities are continuous functions of the external thermodynamic parameters, but their first derivatives and, consequently, the
thermodynamic coefficients show discontinuities. Both from experiments
and theoretical investigations it follows that this classification also does not include all possible cases (ONSAGER, 1944; GEBHARDT, KREY,
1980; ROLOV, JURKEWITSCH, 1983).
The theoretical description of phase transitions is the topic of a
number of investigations. A comprehensive overview on the very different methods applied, the results and applications gives the series of
monographs on the theory of critical phenomena and phase transitions
edited by DOMB, GREEN and LEBOWITZ (1972 - 1983). Remarkable results
in the last decades were obtained e.g. in the description of phase
transitions of second order and critical phenomena (PATASHINSKY,
POKROVSKI, 1982; STANLEY, 1971).
In the same period beyond investigations concerning the further development of the classical nucleation theory (ZETTLEMOYER, 1969, 1977;
HODGSON, 1984; KAGAN, 1960; KOTAKE, GLASS, 1978; SPRINGER, 1978) the
following important contributions to the theory of first-order phase

transitions were delivered:
- the investigation of phase transitions in model systems starting with
  the work of Onsager concerning the theoretical description of the
  Ising model (BINDER, 1977b; PENROSE, BUHAGIAR, 1983; HEERMANN et al.,
  1984)
- the kinetic description of the asymptotic region of the process of
  Ostwald ripening first developed by LIFSHITZ and SLYOZOV (1958, 1959,
  1961), later extended and modified by WAGNER (1961), MARQUSEE and
  ROSS (1983, 1984) and others
- the formulation of a theory of the first stages of spinodal decomposition, of phase transitions starting from thermodynamically unstable
  initial states by HILLERT (1961); CAHN and HILLIARD (1958, 1959);
  CAHN (1961, 1962, 1965, 1966, 1968); COOK (1970); LANGER, BAR-ON and
  MILLER (1975)
- the application of field-theoretical methods to a more precise formulation of nucleation theory and its application to a number of special problems by LANGER (1967, 1969, 1980); LANGER and FISHER (1967);
  TURSKI and LANGER (1980); PATASHINSKY and SHUMILO (1979, 1980), in
  particular, to the problem of nucleation in the vicinity of the critical point
- the development of a qualitatively exact description of the behaviour of real systems in the region of the classical spinodal curve,
  the region between thermodynamically metastable and unstable initial
  states, in terms of a generalized cluster model by BINDER, STAUFFER
  (1976); BINDER (1977a,b); MIROLD and BINDER (1977) and BINDER et al.
  (1978)
- the extension of the theories of phase transitions to two-dimensional
  systems (e.g.: KOCH, DESAI, ABRAHAM, 1983; MARQUSEE, 1984; SAN
  MIGUEL, GRANT, GUNTON, 1985; CAHN, 1977; MILCHEV, MARKOV, 1984;
  CHAKRAVERTY, 1967; GUTZOW, AVRAMOV, 1981).

A further interesting development consists in the investigations of
qualitative changes of the structure or the reaction of nonequilibrium
systems with respect to a variation of the external parameters, their
analogies and differences to equilibrium thermodynamic phase transitions. These effects are closely connected with the development of
dissipative structures (NICOLIS, PRIGOGINE, 1977; EBELING, FEISTEL,
1982; EBELING, KLIMONTOVICH, 1984). These so-called nonequilibrium
phase transitions give the possibility to create structures which are
stable but not in equilibrium with the surrounding medium. Recently
there has been an enormous interest in the theoretical understanding
of domain growth kinetics in quenched nonequilibrium systems under-

going ordering processes. It is believed that the growth of these do-
mains of size l follows power laws $l(t) \sim t^x$ with universal exponents x
(MILCHEV, BINDER, HEERMANN, 1986). Such investigations related to scaling theory are not considered in detail here.

An analysis of the different theoretical approaches to nucleation shows
that in the investigation of nucleation processes it is usually assumed, that the state of the medium does not change during the nucleation
period. Only based on this assumption the derivation of a steady-state
nucleation rate as done in classical nucleation theory, most of its
modifications and generalizations is possible. Such an assumption is
valid for (in the limit) infinite systems and under special thermodynamic constraints. But real systems have a finite size and the formation of the cluster of the new phase leads, in general, to a depletion
of the surrounding the clusters medium.

The investigations presented in this book therefore are devoted to the
problem how the finite size of the system will affect the process of
nucleation and growth of clusters.
For an analysis of this problem we will use mainly macroscopic and, in
particular, thermodynamic concepts. Such an approach guarantees a wide
applicability of the results obtained, since the general thermodynamic
relationships do not depend on the specific properties of the systems
considered. We will assume also that the critical clusters can be described by thermodynamic methods and discuss in detail the problem of a
possible curvature dependence of surface tension for small clusters as
one possibility to obtain correct results for the work of formation of
critical clusters.
In addition to the thermodynamic concept we use deterministic and stochastic approaches in order to describe the kinetics of the evolution
of single clusters respectively cluster ensembles. Our task is to cla-
rify by computer simulations the escape of clusters over an energy barrier. Statistical considerations on the thermodynamic potential of the
cluster distributions complete the given thermodynamic investigations.

In the following we want to point out how the book is organized, gi-
ving a short summary of each chapter.
In Chapter 1 we discuss shortly the experimental situations generating
supersaturated states, the basic assumptions of classical nucleation
theory and the droplet model. The term "cluster" is introduced as a
aggregation of atoms.

In Chapter 2 we start with the analysis of the thermodynamic aspects
of heterogeneous systems and phase transitions in finite systems. Considering the well-known Gibbs' theory and the curvature dependence of
surface tension we go over in Chapter 3 to the problem, how depletion
effects can be taken into account in nucleation theory. As a first
approximation, a quasi-steady-state nucleation rate is formulated,
moreover, it is shown, that thermodynamic investigations give the possibility to develop a general scenario of first-order phase transitions
in finite systems (SCHMELZER, 1985a; SCHWEITZER, 1983; SCHMELZER and
SCHWEITZER, 1985; SCHMELZER and MAHNKE, 1986).
Chapter 4 deals with a stochastic description of nucleation in finite
systems. Based on the calculation of the free energy of the cluster
distribution this more fundamental approach is used to obtain computer
simulations which give a useful demonstration of the earlier stages of
the nucleation process and the time-dependent cluster evolution (EBE-
LING and ULBRICHT, 1986; SCHWEITZER, 1986a).
The assumption of an unchanged state of the surrounding the clusters
medium during the nucleation period leads to the possibility of a se-
parate analysis of nucleation and growth. But in finite systems both
nucleation and growth contribute to the depletion and so both processes have to be considered simulteneously. Therefore in Chapter 5 methods of theoretical description of the growth of single clusters and
ensembles of clusters are summarized. In addition, based on the methods
of thermodynamics of non-reversible processes a deterministic equation
describing the growth of clusters of a new phase in thermodynamic phase
transitions of first order is developed (SCHMELZER, 1985a; SCHMELZER,
ULBRICHT, 1987).
Chapter 6 is devoted to the late stage of tho phase transformation process called Ostwald ripening. Deterministic equations are used to derive a new method of theoretical description of the competitive growth
problem. Both a comparison with the Lifshitz-Slyozov theory is given
and the relations to the theory of selforganizing systems are underlined (SCHMELZER, 1985d; MAHNKE, FEISTEL, 1985; MAHNKE, 1986, 1987).

This general results are applied then (Chapter 7) to the investigation of the growth of single clusters (bubbles, drops, crystallites)
in different systems (SCHMELZER, SCHWEITZER, 1987a). Moreover, it is
shown, that not the specific properties of the systems considered but
the depletion effects lead to the general scenario of the phase transition, discussed in the previously Chapters. It allows a consideration of the influence of elastic strains on the growth of clusters in
solids and the influence of elastic strains on Ostwald ripening. So,
in Chapter 8 first some models are developed for a description of ela-

stic strains evolving due to the formation and growth of clusters
as a basis for a kinetic description of the growth of the clusters in
elastic and viscoelastic media (SCHMELZER, GUTZOW 1986 a, b).

So far we have been dealing with the nucleation of a single vapour
(one-component system), but there are important situations in which
the simultaneous nucleation of two or more substances must be considered. The kinetics of growth of droplets in a binary gaseous mixture
under isothermal, isobaric and adiabatic conditions is described by
SCHMELZER and SCHWEITZER (1987b,d). Binary systems are of interest for
technical applications in the area of energy conversion.
Based on numerical calculations and computer simulations it is shown,
that the process of phase transition in k-component systems (k=1,2,...)
can be divided into three stages, a stage of coupled nucleation and
simultaneous growth of the already formed supercritical clusters, a
second stage of practically independent growth of the clusters their
number being nearly constant and a third stage of competitive growth,
of Ostwald ripening, leading to a decrease of the number of clusters
and an increase of their mean radius.

1.2. Thermodynamic and Experimental Conditions for Supersaturated
      Vapour States

If we consider a vapour at equilibrium, then a certain change of the
thermodynamic constraints (e.g. pressure, temperature, system size)
can remove the system into a nonequilibrium state. The vapour becomes
supersaturated, that means a phase transition is possible leading to
the formation of the liquid phase.
A condensation of the vapour already at its saturation point occurs
only in the presence of sufficient amounts of "foreign" condensation
nuclei, such as particles, ions, and extended surfaces. This type of
condensation is named heterogeneous nucleation. It is a relevant phe-
nomenon e.g. in cloud physics as well as in metallurgical fields
(crystallization in melts,condensation on substrates). However, due to
the complexity of its physical mechanisms the systematic studies of
this problem have little progress up to now. The nucleation on foreign
particles or surfaces is greatly influenced by their geometrical, chemical and physical properties which are extremely varied. Current theories of heterogeneous nucleation are mainly centred around macroscopic

concepts and thus explain macroscopic features of the observed phenomena. Often these investigations are essentially extensions of homogeneous nucleation theories. Indeed,there exist many similarities between
homogeneous and heterogeneous nucleation processes.
We restrict ourselves to the case of homogeneous nucleation when the
vapour is free of all impurities. Homogeneous nucleation seldom occurs
outside of carefully arranged laboratory experiments. Nevertheless, an
understanding of homogeneous nucleation provides considerable insight
into, and appreciation of the more complex heterogeneous nucleation.

In this Chapter first we discuss the thermodynamic conditions to find
a supersaturated vapour. It follows a short review of experimental situations to observe condensation in a supersaturated vapour.
Fig. 1.1 presents the well known p-V diagram of the van-der-Waals gas.
There is indicated the typical van-der-Waals isotherm (a-c-e-g) and
the Maxwell isobar (b-f). The line (j-k-l) gives the saturation line,
also called the binodal. Outside the binodal region only a homogeneous
state exists, but inside the binodal a heterogeneous state, given by
the coexistence of two phases, the vapour and the liquid, is possible.
The line (h-k-i) denotes the spinodal. Inside the spinodal region the
stability condition for the vapour $(\partial p/\partial V)<0$ is not valid and the
vapour phase is an instable state. Immediately a first order phase
transition occurs, the so-called spinodal decomposition where no activation energy is needed for.
The region between the binodal and the spinodal includes the metastable
states, where supersaturated phases are possible to exist over a certain time. the region (j-k-h) represents the superheated liquid, the
region (i-k-l) means the supercooled vapour.

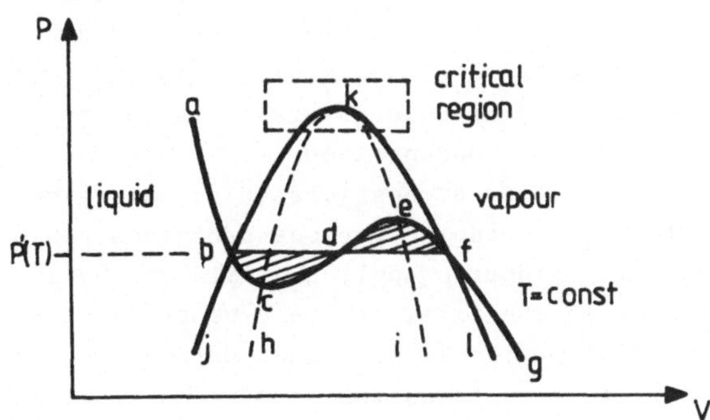

Fig. 1.1:
p-V diagram of a van-der-Waals gas

The Maxwell isobar defines the saturation pressure p' for the given
temperature T, while the pressure of the initially supersaturated vapour is given by p. We now define the initial supersaturation by

$$y_0 = \frac{p}{p'(T)} \tag{1.1}$$

If the supersaturation reaches a critical value, as discussed later,
a phase transition becomes possible in the initially homogeneous metastable vapour. But because now a certain activation energy is required
to insert the formation of the new phase the mechanism of phase transition is quite different from the spinodal decomposition. It is given
by a nucleation process. The nuclei denote small centers of the new
phase which are spatially distributed. The origin of such nuclei is a
random process (discussed in Chapter 4), they must have an overcritical
size to grow further up to a macroscopic liquid phase.
We remark that modern theories often do not divide between nucleation
and spinodal decomposition, because they base on a very general droplet model satisfied in both regions (see e.g. BINDER and STAUFFER,
1976). The classical droplet model used in the following is refered to
the range of low supersaturations far from the critical point.

A thermodynamic analysis of the heterogeneous system consisting of the
nuclei (drops) and the vapour, is given in the Chapters 2 and 3 while
the kinetics of formation and growth of drops will be investigated in
Chapters 4, 5, 6. It is a question of interest to discuss before some
experimental situations where the nucleation and condensation in supersaturated vapours can be observed.
Most of these experiments have been carried out to examine the classical nucleation theory reported in the following Chapter. The results
of the different experiments are summarized in some review articles,
e.g. by SPRINGER (1978), KOTAKE and GLASS (1978), WEGENER and WU
(1977). We neglect here the details of the experiments and restrict
ourselves only to the generation of supersaturated vapour states and
the thermodynamic constraints of condensation.
From a thermodynamic point of view an unsaturated or saturated vapour
may become supersaturated by undergoing various processes, such as (i)
isothermal compression, (ii) isobaric cooling, (iii) isochoric cooling,
(iv) isentropic expansion. In the first of these processes the vapour
temperature remains constant, while it decreases during the latter
processes. Experimentally in most cases the isentropic expansion is
used to generate a supersaturated vapour. Often the vapour is mixed
with a noncondensing (inert) gas which must act as a carrier of the
thermal energy form the latent condensation heat. First experiments on

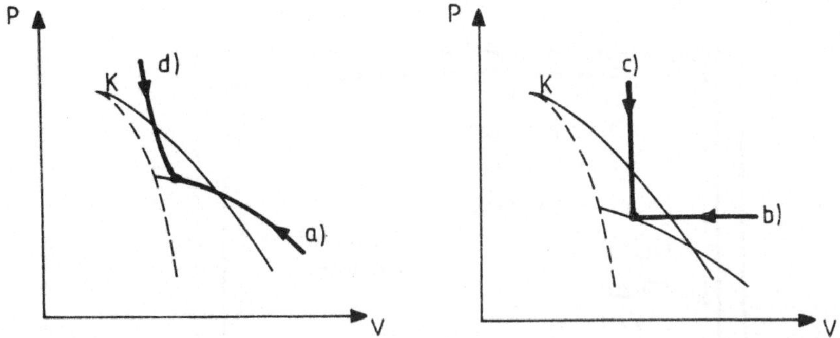

Fig. 1.2:
Generation of a supersaturated vapour by a) isothermal compression,
b) isobaric cooling, c) isochoric cooling and d) isentropic expansion

supersaturated vapours were done with cloud chambers. Expansion cloud
chambers and diffusion cloud chambers are used.
In the expansion chamber the mixture of vapour and inert carrier gas
is expanded rapidly from a known initial volume to a known final volume
under well-controlled conditions. The rapid expansion is usually obtained by actuating a piston or by opening a quick-acting valve to a
lower pressure. Thus the vapour becomes supersaturated. The critical
limit of supersaturation is determined to obtain a remarkable nucleation.
From a historical view Wilson in 1897 performed the first cloud chamber
experiments with water vapour. He found that in a cleaned system no
condensation resulted as long as the supersaturation was less than
about 4. Between a supersaturation between 4 and 8 a small number of
large drops is formed, interpreted by Wilson as condensation on ions.
At supersaturation of 8 and beyond the vapour condensed as a "cloud"
of small water drops. Therefore this critical supersaturation is also
called the "cloud point".
The number of drops is obtained by photographing the drops or from
light scattering techniques.
When the drops are formed the supersaturation is rapidly varying with
time. Its decrease is caused by nucleation and additionally by droplet
growth which occurs simultaneously - a fact not considered in the classical nucleation theory.

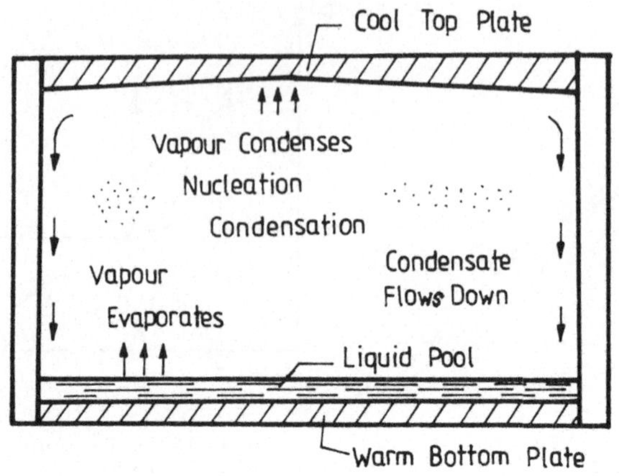

Fig. 1.3:
Sketch of a diffusion cloud chamber
(From: KOTAKE and GLASS, 1978)

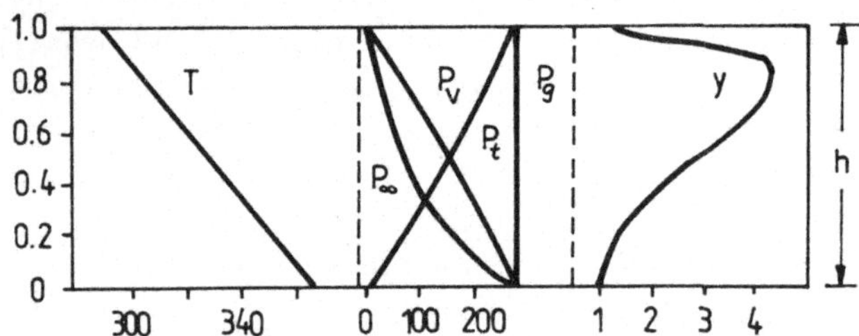

Fig. 1.4:
Change of the thermodynamic parameters in a diffusion cloud chamber
plotted via a reduced chamber hight h
T - temperature (K), p_g - common pressure (Torr), p_v - partial pressure
of the condensable vapour (Torr), p_t - partial pressure of the carrier gas (Torr), p_∞- saturation pressure (Torr), $y = p_v/p_\infty$ - supersaturation
(From: HEIST et al., 1976)

The diffusion cloud chamber is explained in Fig. 1.3. KATZ and co-
workers (1967, 1970, 1975, 1976) have refined the thermal diffusion
cloud chamber to such an extent that it is now one of the best quantitative methods for studying the nucleation process in a gas. The total
density of the vapour-carrier gas-mixture decreases from the bottom to
the top. Since the saturation vapour pressure of the mixture varies
exponentially with temperature and the partial pressure of the vapour

is nearly linear a supersaturated state can be obtained at a certain
hight of the chamber (see Fig. 1.4). The condensation process occurs
for definite thermodynamic conditions. Because the chamber is closed a
stationary nucleation process can be investigated.

Another possibility to study nucleation phenomena is given by nozzle
flows. Here the problem of condensation requires a practical importance for the design and performance of supersonic and hypersonic wind
tunnels, rocket nozzles, steam turbines and cooling towers. When a vapour or a vapour mixture flows along the axis of a supersonic nozzle
(see Fig. 1.5), the vapour pressure decreases with decreasing temperature more rapidly than the expansion pressure down to or below the
saturation pressure. Further expansion may lead to condensation, which
causes first a slight deviation of static pressure from the isentropic
expansion and then an almost abrupt increase in the pressure in a
short distance (Fig. 1.6). In this condensation zone, the maximum heat
release is attained. Further expansion leads to two-phase flows of an

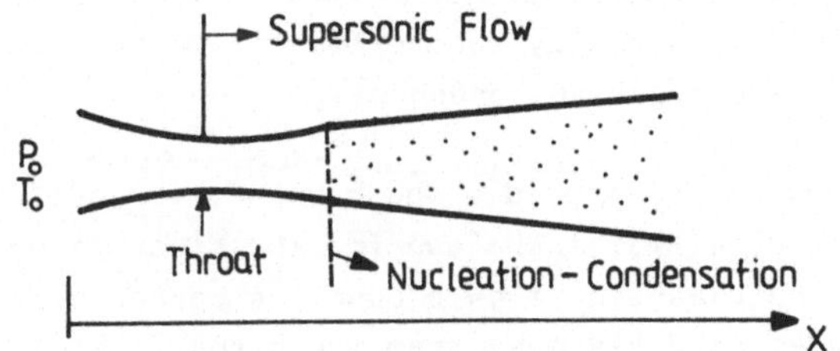

Fig. 1.5:
Sketch of condensation in a supersonic nozzle flow
(From: KOTAKE and GLASS, 1978)

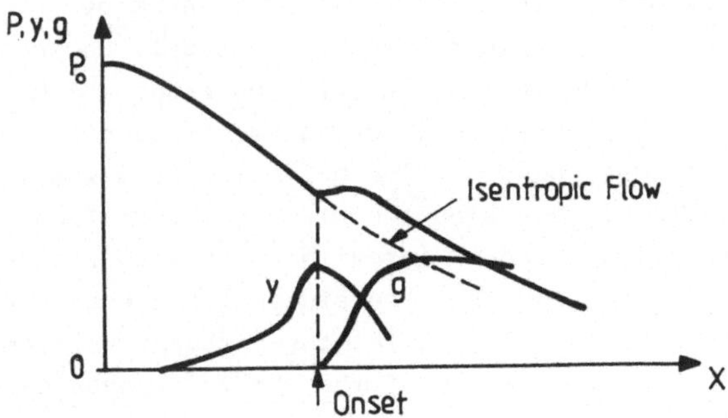

Fig. 1.6:
Change of the thermodynamic parameters in the supersonic nozzle flow
p - vapour pressure; y - supersaturation; g - amount of condensed
matter

asymptotically saturated vapour and liquid droplets. In the late stages the coagulation of the already formed drops has to be considered.

In addition to cloud chamber and nozzle-flow methods, the well established shock-tube technique provides one more possibility for investigating experimentally and analytically nucleation and condensation processes. A shock tube in its simplest form consists of a straight con-
stant-area duct which is divided by a removable diaphragm into a high-
pressure driver section and a low-pressure driven section. By bursting
the diaphragm, a shock wave rapidly develops and moves into the driven
section, and an isentropic rarefaction wave propagates into the driver
section, which will suddenly expand isentropically mixtures of condensing vapour and noncondensable gas. The mixtures are cooled up to or
below the supersaturation state. Further expansion leads to appreciable condensation of the vapour. The onset of condensation can be observed by a sudden pressure rise with respect to the isentropic expansion and/or abrupt increase in the light scattering.
Since the gas dynamic behaviour of such flows is well understood
(WEGENER and WU, 1977) and easily controlled rarefaction waves have
proved most useful investigating condensation.

Steady expansion wave flows around a sharp corner (Prandtl-Meyer expansion) offers an additional technique for the study of nucleation
and condensation. The expansion flow around the corner provides cooling
rates which decreases strongly away from the corner. Hence, the flows
of condensing vapour, the features of nucleation and condensation depend appreciably on the distance from the corner.

The experiments listed above are mainly limited to examine the supersaturation at which appreciable condensation occurs. Consequently,
the details of the collision kinetics to form the drops are hidden. It
is difficult to obtain from macroscopic measurements such as droplet
size, pressure, temperature, density. The molecular beam sampling of
a free jet seems capable of providing the desired information from monomer or dimer collisions to droplet formation. For this reason it has
been used intensively in the past for the study of nucleation kinetics
of early clustering stages. For a free jet the gas expands from the
source into a vacuum. Near the centre of the jet it expands supersoni-
cally as a nearly-isentropic continuum flow until it becomes rarefied
enough to become a free-molecular flow. In the first stage of expansion the cluster kinetics is primarily governed by a single cluster species. Once conditions in the jet favour the growth of this species, the

formation of large clusters will occur without a significant rate of
cluster dissociation.
For an initially supersaturated vapour a finite time is required for
the molecules to achieve appreciable clusterization or critical nucleation. Such relaxation times of nucleation are of the order of micro-
seconds. In contrast to this, in cloud chambers, supersonic nozzles
and shock tubes the thermodynamic state changes sufficiently slowly
compared with the process of clusterization. That's why the kinetically
steady state can be established after a very short induction time for
nucleation and the flow is described by a quasi-steady theory of nucleation. Only under rapid expansions of free molecular beams into
higth vacuum the quasi-steady assumption of clusterization is no longer valid and a kinetic treatment must be used. But for the kinetic
process it is difficult to establish a quantitative scheme without
prior knowledge of the rate of the clustering process.

In the following the quasi-steady theory of nucleation shall be outlined in terms of the classical nucleation theory.

1.3. Outline of Classical Nucleation Theory

1.3.1. The Classical Droplet Model

The basic ideas of the classical nucleation theory were developed
half a century ago by VOLMER and WEBER (1926), BECKER and DÖRING (1935),
ZELDOVITSCH (1942) and FRENKEL (1955) and others. More recent surveys
of this area including new aspects of the theory are presented by
HIRTH and POUND (1963), ZETTLEMOYER (1969, 1977), ABRAHAM (1974) (see
also: NISHIOKA and POUND, 1977; SPRINGER, 1978; KOTAKE and GLASS,
1978).
The classical nucleation theory with its modifications which will be
shortly discussed in the following is up to date one of the most applicable theories to explain experimental results, and it agrees in a
certain range of deviation with the experiment. Despite its practical
importance a number of problems arise caused by a rather phenomenological derivation of the theory. To avoid these difficulties new theories have been established, we note e.g. the improvement of the classical cluster approach by BINDER and STAUFFER (1976) or the field
theoretic approach on nucleation theory.

The last decade has been an explosive burst of activity in studies of
clusters with emphasis on their dynamics of formation, chemical properties, structure and stability. It is useful to provide a classification
of clusters according to size n. Clusters are comprised of an aggregation of components, the smallest constituent is the dimer with n = 2
monomers. Then, in order of increasing size, we distinguish between
microclusters (n = 2 - 10), small clusters (n ~10 - 100) and large
clusters (n > 10^2) (JORTNER, 1984; compare also Fig. 4.2). This level
structure indicates that characteristics of clusters differ from the
properties of bulk material. The binding energies of atoms in clusters
or drops are a function of size n.
The formation of rare-gas clusters is studied experimentally and theo-
retically, so the clustering in van-der-Waals systems is of interest
(HEYES, WOODCOCK, 1986). Another direction of research is connected
with the properties of ionic clusters (CASTLEMAN, KEESEE, 1986). The
formation of ion clusters in a gas phase proceeds via a sequence of
association reactions between ions and the molecules with which they
interact.
The classical nucleation theory is based on the major and rather simple
droplet model which presumes an incompressible cluster being at rest.
The surface of this cluster can be described by the macroscopic flat-
film surface tension $\mathcal{G}$ (the so-called capillarity approximation), the
density is given by the macroscopic liquid density ϱ_α (see Fig. 1.7).

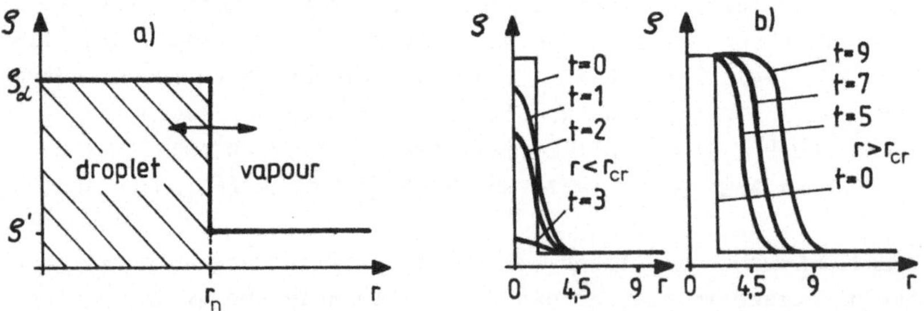

Fig. 1.7:
(a) Classical droplet model. ϱ_α denotes the macroscopic liquid density,
 $\mathcal{G}'$ the density of the saturated vapour
(From: ULBRICHT et al., 1986)
(b) Density profiles of a shrinking (undercritical) droplet in comparison with a growing (overcritical) drop.
(From: EBELING and FEISTEL, 1982)

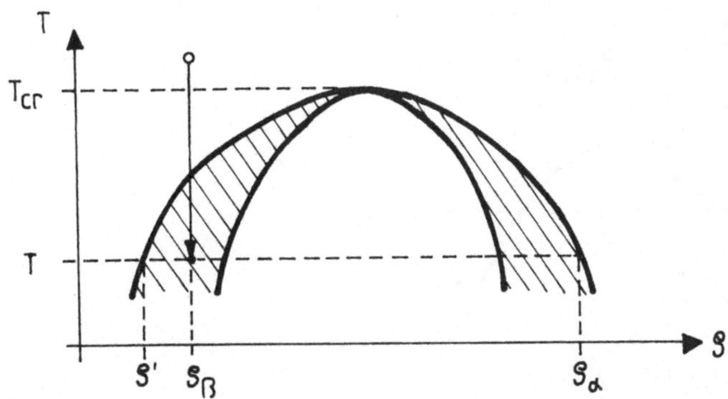

Fig. 1.8:
Phase diagram of the vapour-liquid system. The system is quenched from the one-phase equilibrium to a supersaturated (metastable) state in the binodal region. ϱ_β is the vapour density of the supersaturated state, ϱ' the equilibrium vapour density, ϱ_α the liquid density at the given temperature T. T_{cr} denotes the critical temperature.

The difficulties of this approach become evident in the range of small cluster sizes, because the extrapolation of the macroscopic density and the surface tension to small clusters fails.

The free energy to form a drop with n particles comprises two parts:

$$\Delta F_n = (\mu_\alpha - \mu_\beta)\, V_n (\varrho_\alpha - \varrho_\beta) + \sigma A_n \tag{1.2}$$

The first term of the right hand side of eq. (1.2) is the volume free energy of the drop and represents the decrease in chemical potential due to the formation of the drop. The second term is the contribution of the surface free energy and represents an increase in ΔF_n. V_n is the volume of the drop, A_n its surface, μ_α denotes the chemical potential of the drop, ϱ_α its density, while μ_β stands for the chemical potential of the surrounding vapour and ϱ_β for the density of the supersaturated vapour phase (compare Fig. 1.8).

In the classical theory μ_β and ϱ_β are assumed to be constant, that means a change of the vapour state due to the formation of the drop is not considered.

Additionally, the assumptions are made that the vapour is ideal, the latent heat released at the droplet-vapour- interface is negligible, and the nucleation is isothermal.

For an ideal vapour and an incompressible liquid it yields:

$$\mu_\alpha - \mu_\beta = - k_B T \ln \varrho_\beta / \varrho' = - k_B T \ln y$$

The ratio ϱ_β / ϱ' gives the supersaturation already introduced in eq. (1.1). It is a constant because of ϱ_β = const.

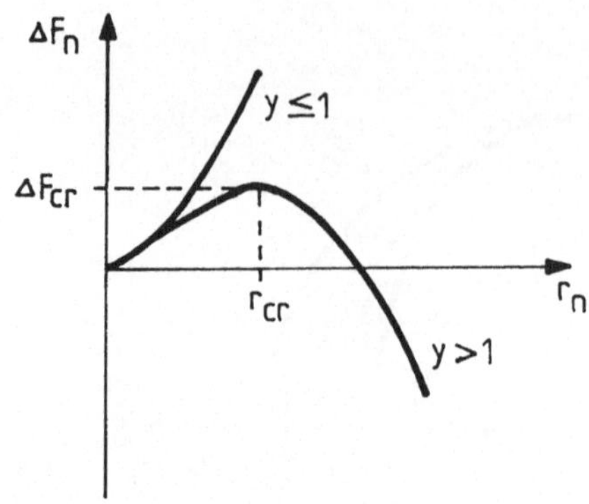

Fig. 1.9:
Sketch of the formation energy ΔF_n (eq. (1.3)) versus droplet radius r_n. y means the supersaturation. The critical droplet size is given by eq. (1.4), the maximum of the free energy by eq.(1.5).

Neglecting further ϱ_β in comparison with ϱ_α and assuming a spherical drop with the radius r, we obtain from eq. (1.2):

$$\Delta F_n = - \frac{4\pi}{3} r_n^3 \, k_B T \, \ln y \; + 4\pi\sigma \, r_n^2 \tag{1.3}$$

The free energy of the droplet formation is plotted in Fig. 1.9 versus droplet radius r. For initial supersaturations $y \le 1$ the energy is a monotonously increasing function of the radius, that means the vapour will be the absolutely stable phase. For $y > 1$ ΔF_n decreases continuously, when the drop has reached an overcritical size $r > r_{cr}$. This critical radius follows from the extremum condition $(\partial \Delta F_n / \partial r_n)=0$:

$$r_{cr} = \frac{2\sigma}{\varrho_\alpha k_B T \, \ln y} \tag{1.4}$$

With eq. (1.4) it yields for the maximum of the formation energy:

$$\Delta F_{cr} = \frac{1}{3}\sigma(4\pi r_{cr}^2) = \frac{16\pi\sigma^3}{3(\varrho_\alpha k_B T \, \ln y)^2} \tag{1.5}$$

Since ΔF has a maximum the equilibrium between the drop and the surrounded vapour is an instable one, as pointed out in detail in Chapter 3.
The maximum of ΔF_n is related to a nucleation barrier which the drop must cross over to reach a macroscopic size. Therefore overcritical drops can grow further while undercritical drops have to diminish again in a deterministic sense.

1.3.2. Kinetic Assumptions of Classical Nucleation Theory

A complete description of nucleation requires information about the
kinetics of each individual forward and reverse step in the sequence
of clustering reactions. At present, there is no general theory of
association reactions which can be used to treat many-body interactions
responsible for the formation of a critical-size nucleus.
The classical nucleation theory assumes that the formation of drops
takes place by progressive aggregation of only single vapour molecules
A_1 (also called monomers). The formation can be represented by the
"reaction"

$$A_n + A_1 \underset{e_{n+1}}{\overset{c_n}{\rightleftharpoons}} A_{n+1} \tag{1.6}$$

The drop may also dissociate by evaporation of monomers, as indicated
by the arrows. c_n means the condensation rate at the surface of the
drop and e_n the evaporation rate. Classically the evaporation rate is
assumed to be constant and not influenced by the conditions in the vapour, while the condensation rate is chosen proportional to the surface of the drop and to the pressure of the supersaturated vapour.

The kinetic ansatz (1.6) does not consider collisions between more
than two particles, it also neglects the collisions between two clusters of larger sizes and the break-off of a drop into several pieces.
Obviously, in the ansatz (1.6) the clusters and their reactions are
not considered in all details (e.g. possible atomic arrangement), but
rather as effective clusters where such geometric details are averaged
out.
A change in the number of drops consisting of n particles can occur in
four variants. The number N_n of drops of size n will increase due to
the addition of the monomer to a drop of size (n-1) and due to the
evaporation of one molecule from a drop of size (n+1). N_n will decrease
due to condensation on or evaporation from a drop of size n. The rate
of change in N_n can be represented therefore by a continuity equation
of the form:

$$\frac{\partial N_n}{\partial t} = c_{n-1}N_{n-1} + e_{n+1}N_{n+1} - (c_n + e_n)\, N_n \tag{1.7}$$

By defining a net rate

$$I_n = c_n N_n - e_{n+1}N_{n+1} \tag{1.8}$$

we can write eq. (1.7) as:

$$\frac{\partial N_n}{\partial t} = I_{n-1} - I_n \quad . \tag{1.9}$$

The net rate I_n is time dependent and can be viewed as a current of drops flowing from the state n to the state n+1. It is interpreted as the nucleation rate to form drops of size $n+1$.

Eq. (1.7) respectively (1.9) is valid under both unsteady and steady-state conditions. A main task of the classical nucleation theory is the calculation of the steady state nucleation rate I^S. It follows from the stationary solution of eq. (1.9). The condition $\partial N_n / \partial t = 0$ leads to:

$$I_{n-1} = I_n = I_{n+1} = \ldots = I^S = \text{const} \tag{1.10}$$

with

$$I^S = c_n N_n^S - e_{n+1} N_{n+1}^S \tag{1.11}$$

N_n^S means the stationary droplet distribution. To obtain the known form of the classical nucleation rate we rewrite eq. (1.11) in the form:

$$I^S = c_n N_n^O \left(\frac{N_n^S}{N_n^O} - \frac{e_{n+1}}{c_n} \frac{N_{n+1}^S}{N_n^O} \right) \tag{1.12}$$

We have introduced here the equilibrium droplet distribution N_n^O which must validate the condition that the current I vanishes in thermodynamic equilibrium. From the condition $I = 0$ we get:

$$c_n N_n^O - e_{n+1} N_{n+1}^O = 0 \tag{1.13}$$

Eq. (1.12) means a system of equations valid for all values of n. Summing these equations up to a number $n^* > n_{cr}$ and using eq. (1.13) we obtain:

$$\sum_{n=1}^{n^*} \frac{I^S}{c_n N_n^O} = \frac{N_1^S}{N_1^O} - \frac{N_{n^*}^S}{N_{n^*}^O} \tag{1.14}$$

Classically this equation has to fulfil two boundary conditions:
(i) All supercritical clusters of a size $n \geqslant n^*$ are instantaneously removed from the system that means $N_n = 0$ for $n \geqslant n^*$. In order to keep the total number of particles constant, we add to the system the same number of single particles as have been removed with the droplets, that are $n^* N_{n^*}^S$ particles. Thus the stationary process will not be troubled.
(ii) The stationary distribution N_n^S of small drops can be approximated by the equilibrium distribution N_n^O:

$$\lim_{n \to 1} \frac{N_n^s}{N_n^o} = 1$$

By use of these boundary conditions eq. (1.14) yields:

$$I^s = \left\{ \sum_{n=1}^{n^*} \frac{1}{c_n N_n^o} \right\}^{-1} \qquad (1.15)$$

Now it is necessary to know the equilibrium cluster distribution N_n^o. From thermodynamic investigations (FRENKEL, 1955) it is received as follows

$$N_n^o = N \exp\left\{ - \frac{\Delta F_n}{k_B T} \right\} \qquad (1.16)$$

ΔF_n gives the change of the free energy when a drop of size n has been formed (eq. (1.2) or (1.3)).

To find a closed form of the solution (1.15) some suitable approximations are introduced not discussed here in detail. For instance eq. (1.15) is replaced by an integral equation. Further, the free energy in eq. (1.16) is expanded in terms of a Taylor series about the critical droplet size n_{cr}, where ΔF_n has a maximum. Indeed, it can be shown (BECKER and DÖRING, 1935) that the remarkable contributions to the steady state nucleation rate arise from the surroundings of the critical state. Last but not least we note the approximation of the condensation rate being constant.

Finally the following expression for the classical nucleation rate is obtained (FRENKEL, 1955):

$$\bar{I}^s = c_{cr} \varkappa \mathcal{N}_{cr}^o \qquad (1.17)$$

with

$$c_{cr} = \eta_o \frac{4 \pi r_{cr}^2 \, p}{(2\pi m k_B T)^{1/2}} \qquad (1.18)$$

$$\varkappa = - \frac{1}{2\pi k_B T} \left. \frac{\partial^2 \Delta F_n}{\partial n^2} \right|_{n=n_{cr}} = \sqrt{\frac{\sigma}{k_B T}} \, \frac{1}{2\pi r_{cr}^2} \qquad (1.19)$$

$$\mathcal{N}_{cr}^o = \frac{N}{V} \exp\left\{ - \frac{\Delta F_{cr}}{k_B T} \right\} = \frac{p}{k_B T} \exp\left\{ - \frac{4\pi}{3} \frac{\sigma}{k_B T} r_{cr}^2 \right\} \qquad (1.20)$$

Tab. 1.1:
Critical condensation rate (eq. (1.18)), Zeldovitsch factor (eq. (1.19)), critical droplet radius (eq. (1.4)), maximum free energy (eq. (1.5)), steady state nucleation rate (eq. (1.17)) for different super-saturations, vapour: ethanol, T = 290 K

y	c_{cr} $(10^{10}\ s^{-1})$	$\varkappa$ (10^{-2})	r_{cr} $(10^{-9}\ m)$	ΔF_{cr} $(k_B T)$	$\bar{I}^S$ $(s^{-1}m^{-3})$
1.7	2.94	0.9	2.01	95.4	10^{-8}
1.8	2.54	1.0	1.81	77.8	$9\cdot10^{-2}$
1.85	2.38	1.1	1.73	71.0	88.2
1.9	2.25	1.2	1.66	65.2	$3\cdot10^4$
2.0	2.03	1.5	1.54	55.9	$3\cdot10^8$
2.1	1.86	1.7	1.43	48.8	$5\cdot10^{11}$
2.4	1.52	2.4	1.21	35.1	$6\cdot10^{17}$

(From: SCHWEITZER, 1986a)

$\bar{I}^S$ expresses the number of droplets per unit time and unit volume which reach the critical state in a stationary regime. $\varkappa$ is called the Zeldovitsch nonequilibrium factor (ZELDOVITSCH, 1942) which accumula-tes the difference between the steady-state concentration and the equi-librium concentration of droplets. Its influence on the nucleation ra-te is nearly insignificant. The influence of the supersaturation y on the nucleation rate is more impcrtant while y enters also in the cri-tical radius (eq. (1.4)). Combining the preexponential factors eq. (1.17) reads in the final form

$$\bar{I}^S = \bar{I}_0 \exp\left\{ -\frac{\Delta F_{cr}}{k_B T} \right\} \qquad\qquad (1.21)$$

where $\bar{I}_0$ is of order $10^{31}\ s^{-1}m^{-3}$.
The steady state nucleation rate and the variables of interest are cal-culated in Tab. 1.1 for different supersaturations. The strong depen-dence of the nucleation rate on the supersaturation is also shown in Fig. 1.10, first given by Becker and Döring in 1935.
The supersaturation at which an appreciable formation of nuclei occurs is the already mentioned cloud point. Because $\bar{I}^S$ increases enormously with increasing supersaturation, it is nearly equal to choose $\bar{I}^s = 10^2\ s^{-1}m^{-3}$ or $10^4\ s^{-1}m^{-3}$ as a remarkable nucleation.
We denote that other classical nucleation rates have been derived which differ something from Frenkels expression (eq. (1.17)), but this deviations are not significant as has been pointed out (SPRINGER, 1978).

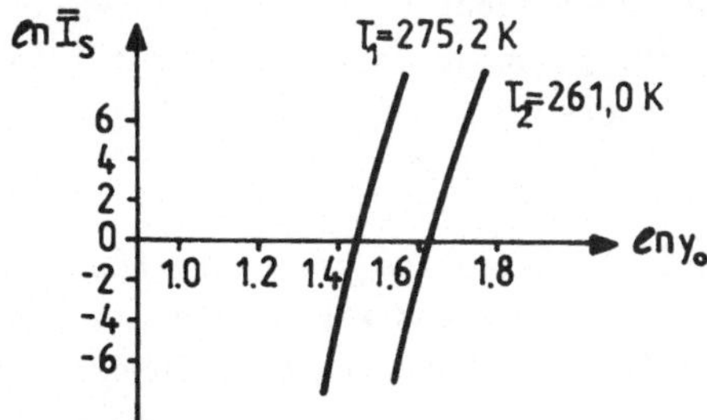

Fig. 1.10:
Nucleation rate as function of supersaturation for two temperatures,
calculated for water
(From: BECKER and DÖRING, 1935)

1.3.3. Modifications of Classical Nucleation Theory

The classical nucleation theory discussed above is based on the following assumptions, which are summarized here again:
- the surface energy of a small drop can be described by the flat-film
  surface tension (so-called capillary   approximation)
- the features of the condensation process are considered in terms of
  steady-state kinetics
- the clusters are at "rest" undergoing neither translational or rotational motion
- the vapour phase is a perfect gas with a constant pressure, that
  means the formation of clusters do not change the vapour state
- the latent heat released at the cluster-vapour interface due to condensation is negligible and the nucleation is an isothermal process.

Many efforts have been made to eliminate some of these assumptions
from the classical theory. We review here only some of these ideas. A
number of problems in the classical nucleation theory arise from the
imprecise concept of a cluster which is rather phenomenologically derived. Therefore methods of statistical mechanics were applied to
avoid some problems. A completely microscopic theory of nucleation was
proposed by KOBRAEI (1984). Unfortunately, it is difficult to obtain
numerical results from the statistical mechanical approach and to re-
late the results to experimentally measurable parameters. A possible
expansion of the classical nucleation theory is the compromise to in-
corporate into the macroscopic thermodynamic approach of the classical

theory corrections obtained by the methods of statistical mechanics.
We now discuss some of the points given above to improve the basic
ideas of classical nucleation theory.

(i) Surface energy

Methods of statistical mechanics to avoid the capillary approximation
were applied by ABRAHAM (1982),ROWLINSON and WIDOM (1982) and others.
The problem of surface tension is not considered here in detail because
an examination of the flat-film surface tension applied to small clusters will provided in Chapter 2. We want to underline here only the
importance of a correct value of the surface tension σ: The steady-
state nucleation rate (eq. (1.21)) depends very strongly on σ, because
σ^3 enters the exponential of eq. (1.21). That's why a change in σ of
only 10% leads to a change in the nucleation rate by eight orders of
magnitude (compare Tab. 1.2).

Tab. 1.2:
Critical droplet radius (eq. (1.4)), maximum free energy (eq. (1.5))
and steady-state nucleation rate (eq. (1.21)) in dependence on a slightly changing value of the surface tension σ_0. vapour: ethanol, supersaturation y = 2, T = 290 K

	σ $(10^{-2}$ N/m)	r_{cr} $(10^{-9}$ m)	ΔF_{cr} $(k_B T)$	I^s $(s^{-1}$ m$^{-3})$
σ_0 - 10%	2.028	1.385	40.72	$2.8 \cdot 10^0$
σ_0	2.254	1.541	55.91	$3.0 \cdot 10^8$
σ_0 + 10%	2.479	1.693	74.39	$1.1 \cdot 10^{15}$

(ii) Nonstationary nucleation rate

The steady-state nucleation rate is calculated for the condition that
the size distribution of clusters does not change in time. This assumption fails mainly in two cases: At the very beginning of the nucleation
process a certain time is required to establish the steady-state distribution, that is the so-called time lag. During this transient time
the nucleation rate reaches its stationary value. On the other hand,
in some experiments, e.g. for a rapid expansion in supersonic free molecular jets, the relaxation process into the steady-state takes a much
longer time than the characteristic lifetime of the supersaturated vapour system. That means a steady-state nucleation does not exist. For
this purpose, non-steady kinetics of clusterization must be considered.
As already mentioned the rate equation (1.7) is valid both for steady
and nonsteady situations. In the latter case a system of equations has
to be solved numerically for every time step. (The number of equations

agrees with the number of possible cluster sizes.)
We restrict ourselves here to the case that a steady-state solution
of the rate equation exists and discuss the time dependent change of
the nucleation rate to reach the steady value (see e.g. KATZ and
WIEDERSICH, 1977; KATZ and DONOHUE; 1979; KASHCHIEV, 1969; KELTON et
al., 1983). From analytical calculations it is found:

$$\bar{I}(t) = \bar{I}^S \, (1 - \exp(-\,t/\vartheta)) \tag{1.22}$$

with $\bar{I}^S$ being the steady-state nucleation rate (eq. (1.21)) and ϑ
being the time lag to reach the steady-state. ϑ is obtained in the
form:

$$\vartheta = C/(c_{cr}\,\varkappa^2) \tag{1.23}$$

$\varkappa$ is the Zeldowitsch factor (eq. (1.19)) and c_{cr} the condensation rate
at the critical cluster (eq. (1.18)). C means a constant which is calculated somthing different by different authors. The values change
from 0.03 to 0.5 (SPRINGER, 1978).
A well established expression for the nonstationary nucleation rate
was given by KASHCHIEV (1969), which agrees also with the results of
computer calculations by KELTON et al. (1983). He found:

$$\bar{I}(t) = \bar{I}^S \left[1 + 2 \sum_{m=1}^{\infty} (-1)^m \exp\left(-\,\frac{m^2\,t}{\vartheta}\right) \right] \tag{1.24}$$

with C = 0.12 (eq. (1.23)).
Fig. 1.11 presents the nonstationary nucleation rate and the number
of overcritical clusters as a function of time. The effective time lag
θ approximates the time to establish the steady-state distribution of
critical clusters.

(iii) Consideration of droplet motion
The classical droplet model assumes the cluster to be at rest. But in
general, the clusters in the vapour may behave dynamically like large
gas molecules, having free translational and rotational motions. An
approach to consider these motions was first given by POUND and co-
workers (LOTHE and POUND, 1966; ABRAHAM and POUND, 1968) and by Reiss
and coworkers (REISS and KATZ, 1967; REISS et al., 1968). It assumes
basically a droplet model with statistical mechanical corrections.
These corrections are expressed in terms of the translational and rotational partition functions, q_{trans} and q_{rot}, treating the cluster
as an rigid body. The six degrees of freedom used in q_{trans} and q_{rot}
must compensate by introducing a "replacement factor" q_{rep} which is
related to the internal partition function of the cluster.

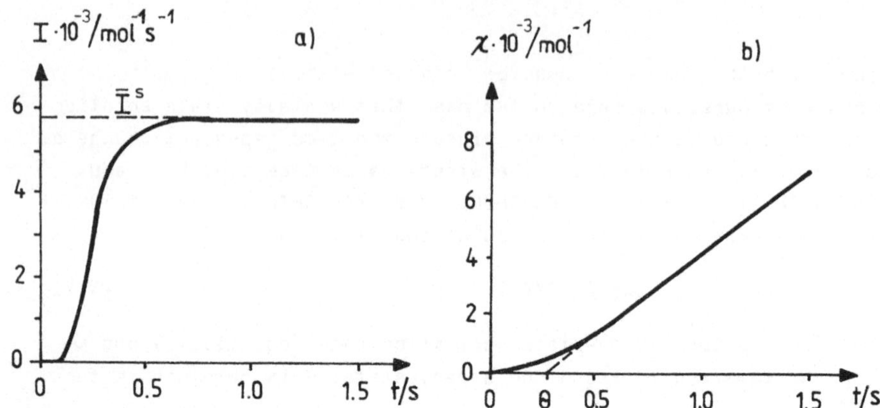

Fig. 1.11:
The nucleation rate I for critical clusters (a) and the number of overcritical clusters χ (b) as a function of time. $\bar{I}^s$ denotes the steady-state nucleation rate. The effective time lag θ to establish the steady-state distribution of critical clusters is marked

(From: KELTON et al., 1983)

From the mentioned degrees of freedom additional contributions to the free energy of droplet formation arise, which decrease the value of the free energy significantly, for water vapour by about -40 $k_B T$ as it was pointed out by Lothe and Pound: The translational contribution to the free energy is of the order -24 $k_B T$, the rotational contribution of the order -21 $k_B T$ respectively. From the replacement term a value of about 5 $k_B T$ is obtained.

The additional contribution to the formation energy result in a correction of the critical cluster size due to the droplet motion. Instead of the classical equation (1.4) it yields

$$r_{cr} = \frac{2\sigma}{\rho_\alpha k_B T \ln y} \left(1 - \frac{3 k_B T}{2\pi\sigma r_{cr}^2} \right) \tag{1.25}$$

which means a decrease of the critical droplet size.

The expression of the classical nucleation rate $\bar{I}_{cl}^s$ (eq. (1.21)) must be corrected as follows:

$$\bar{I} = \frac{q_{trans} \, q_{rot}}{q_{rep}} \; \bar{I}_{cl}^s \tag{1.26}$$

LOTHE and POUND (1966) estimated the pre-factor $q_{trans} q_{rot}/q_{rep} \approx 10^{17}$ for water vapour. From something different application of statistical mechanics to the droplet motion problem (REISS et al. (1968)) already obtained an increase in the nucleation rate of 10^3 to 10^6.

The significant disagreement of the Lothe-Pound theory with experiments leads to a broad discussion of the so-called "translation-rotation paradox" (REISS and KATZ, 1967) which cannot reviewed in detail here (see e.g. ABRAHAM, 1974; NISHIOKA and POUND, 1977). Despite extensive discussions appeared on this matter the problem is not completely solved today.

(iv) Finally, we give some short remarks on additional efforts to improve the classical nucleation theory. The effects of vapour phase nonideality on nucleation were investigated e.g. by KATZ and BLANDER (1973), while LOVETT (1984) investigated the influence of hydrodynamic properties of the system on the nucleation rate. A theory of nucleation in associated vapours was pointed out by KATZ, HEIST and others.

   The latent heat from the condensation has to consider by the nucleation theory if the relaxation processes occur slowly. Correction factors for the non-isothermal removal of latent heat were proposed by LUDIG (1975) and BARSCHDORFF (1975).
Classical nucleation theory breaks down near the spinodal and near the critical point because the correlation length becomes of the order of the system size. In this case new theories have been developed to describe the behaviour of the system during the phase transition. Some investigations we mentioned already in Chapter 1.1.

1.4. Nucleation in a Lattice Gas Model

The theoretical description of the nucleation process in a real system, o.g. a supersaturated vapour, still encounters a number of difficulties, as it has been shown in the previous section. Therefore, for a detailed understanding of phase separation often simple model systems are used which allow a deep insight into the elementary processes. The evolution of the system can be studied in terms of computer experiments. Although they are based on a very simplified version of reality, these experiments have some advantages. Because the term cluster has an unambiguous meaning, it is rather straightforward to study cluster reactions (KERLEY, 1986). Also for small densities of a certain type of atoms the size distribution of clusters of such atoms can be observed. Thus the computer simulation technique which is close related to the models considered provides a test of the general features of nucleation theory.

Now we consider a simple model system wherein all details of the nucleation may be observed on a microscopic scale both in space and time. This system is the kinetic Ising or lattice gas model where the atoms may occupy the sites of a regular lattice, with one atom per site only. Denoting the sites by i (i=1,...,N) it yields c_i = 1 if the atom has been occupied this site and c_i = 0 otherwise.

Assuming in contrast a binary alloy each lattice site is occupied by either an A-atom (c_i^A = 1) or B-atom (c_i^B = 1). This model can be used to study decomposition in alloys. The states change again according to a Markov process (see also Chapter 4) where the allowed transition are interchanges of the atoms on two neighbour sites. It can be shown in a simple way that this model is mathematically equivalent to the lattice gas model with a nearest-neighbour interaction.

An appropriate change of the chemical potential brings the system into an unstable state from which the stable one may be reached via nucleation processes. The atoms are removed from (or put on) the sites in a random manner but in such a way that thermal equilibrium is acieved.

We give now only a short review on the kinetics of the lattice gas model, for a detailed explanation see BINDER (1977b), PENROSE and BUHAGIAR (1983), HEERMANN et al. (1984), LEBOWITZ et al. (1982) and literature cited therein.

Following BINDER (1977b) a basic feature of the interactions between the atoms in a lattice gas is that the forces are attractive if the atoms are not too close together. In the opposite case the forces are strongly repulsive. The requirement that each size of the regular lattice can be occupied only once accounts for the hard-core repulsion. Assuming pairwise interactions $\Phi(r_i,r_j)$ the Hamiltonian may be written as

$$H = \sum_{i \neq j} \sum c_i c_j \Phi(r_i,r_j) + \sum_i c_i g(r_i) + H_o \qquad (1.27)$$

where the sums extend over all N sites of the lattice, c_i=1 if the lattice site r_i is occupied by an atom and zero otherwise. $g(r_i)$ is the chemical potential which is allowed to be inhomogeneous. H_o may be represented the independent Hamiltonian of the underlying lattice, which contains the kinetic energy of the atoms. Eq. (1.27) can be transformed into the Hamiltonian of an anisotropic spin one half ferromagnetic Ising model (BINDER, 1977b) which is usually discussed. In this case $\mu_i = \pm 1$ gives the magnetic moment at a lattice site r_i.

Now we define the order parameter $\langle\mu\rangle$ by: $\langle\mu\rangle = 1/N \sum \mu_i$. It is

$$\langle\mu\rangle = \frac{1}{N} \sum_i \left(\langle c_i^A \rangle - \langle c_i^B \rangle \right)$$

in the case of a binary alloy, or

$$\langle\mu\rangle = \frac{1}{N} \sum_i \left(1 - 2 \langle c_i \rangle \right)$$

in the case of the lattice gas model underlying eq. (1.27).
The lattice gas model exhibits a critical temperature T_c similar to the
gas-liquid critical temperature. For an always homogeneous system per-
fect order is reached, that means $\langle\mu\rangle = +1$ or $\langle\mu\rangle = -1$. But below T_c
this order will be imperfect and "clusters" of reversed spins will
appear on the background of the otherwise homogeneously magnetized
phase (or clusters of B atoms on a background of A phase, respectively). An unambiguous meaning of the term "cluster" is obtained if it is
defined as a group of spins linked together by nearest neighbour bonds
in the lattice. Let us note, that in the case of nearest-neighbour interactions there would be no direct interactions between the various
clusters present in the system.
Denoting the time dependent concentration of clusters of size k by $n_k(t)$
for the time dependence of the order parameter the relation

$$\langle\mu(t)\rangle = 1 - 2 \sum_{k=1}^{\infty} k\, n_k(t) \tag{1.28}$$

is obtained (BINDER, 1977b). n_k is normalized so that for a complete
change of the magnetic moments from the initial state (+1) to (-1) the
sum $\sum k n_k$ is just equal to one.
Assuming the system with a perfect order in the initial state, at the
time t = 0 a sudden change of the external variables is performed, e.g.
the magnetic field is changed by a value h, which measures the distance to the coexistence curve h = 0. h is related to the supersaturation
of a vapour state. In dependence on the value of h the system becomes
unstable and the formation of clusters occurs. As it is also known
from vapour experiments, the nucleation rate will be extremely small,
if the supersaturation is sufficiently small, thus the system will remain in a metastable state for a rather long time. Metastable states
in a lattice gas model and their lifetimes have been investigated by
BINDER (1977b).
The kinetics of the simple spin-flip Ising model considered here can
be described by a Master equation (compare Chapter 4). Denoting the
configuration of all magnetic moments μ_i by the vector $\underline{X} = (\mu_1 \mu_2 \ldots \mu_N)$
the probability $P(\underline{X},t)$ to find a given configuration at the time t

changes with time as follows:

$$\frac{\partial P(\underline{X},t)}{\partial t} = \sum_{\underline{X}'} w(\underline{X}' \rightarrow \underline{X})P(\underline{X}',t) - \sum_{\underline{X}'} w(\underline{X} \rightarrow \underline{X}')P(\underline{X},t) \tag{1.29}$$

$w(\underline{X} \rightarrow \underline{X}')$ is the probability that a transition from $\underline{X}$ to $\underline{X}'$ occurs per unit time. Eq. (1.29) has the structure of a rate equation describing the balance between all processes where the system is flipped out and flipped into the considered state $\underline{X}$.

If we consider only those transitions which permit the interchange of a spin μ_i with a neighbouring spin μ_j (Kawasaki dynamics) the transition probability can be chosen in the form (GUNTON et al., 1983):

$$w(\mu_i \rightarrow -\mu_i) = \frac{1}{2\tau}\left\{1 - \tanh\left(\delta H_i/2\ k_B T\right)\right\} \tag{1.30}$$

Here δH_i is the change in (magnetic) energy produced by the spin flip at site i which can be easy calculated from the Hamiltonian of the Ising system. The arbitrary simple spin-flip time τ fixes the time scale. We note, that the transition probability (eq. (1.30)) satisfies the condition of detailed balance.

The calculations described in the following are restricted to a L x L square lattice with nearest-neighbour interactions. The computer simulations were done with the Monte Carlo algorithm (BINDER, 1979, 1984). Fig. 1.12 presents some snapshot pictures for different times after the change h of the field; in Fig. 1.13 the corresponding development of the time dependent order parameter is shown.

The considered system was in thermal equilibrium at time t = 0, before the switch of the field h. One sees from $\langle \mu\ (t=0) \rangle \neq 1$ (Fig. 1.13) that already in an equilibrium state there is a tendency of the atoms to occupy lattice sites and to build up small clusters. This cluster pattern changes rapidly with time due to spin-flips, that means the appearance or disappearance of single atoms at lattice sites. Some of the clusters will grow, others will shrink or disappear. It is also possible to observe cluster reactions. Clusters can move together and form bridges and then building up a rather large cluster, which quickly increases.

The pictures of Fig. 1.12 give only some qualitative impressions on the nucleation process. But it is possible to get from the Monte Carlo simulations also quantitative informations about the total energy or the cluster distribution or relaxation times and so on, which will be not discussed furthermore.

The computer simulation technique applied to Ising models was also used to examine the predictions of classical nucleation theory. As it has

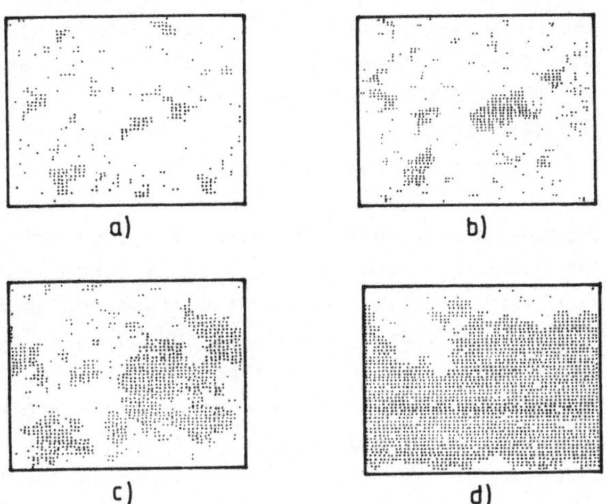

Fig. 1.12:
Snapshot pictures of a 55 x 55 Ising lattice at different times
a) t = 15, b) t = 50, c) t = 70, d) t = 145 (in Monte Carlo steps/
spin)
Atoms accupying lattice sites are indicated by black dots

(From: BINDER, 1977b)

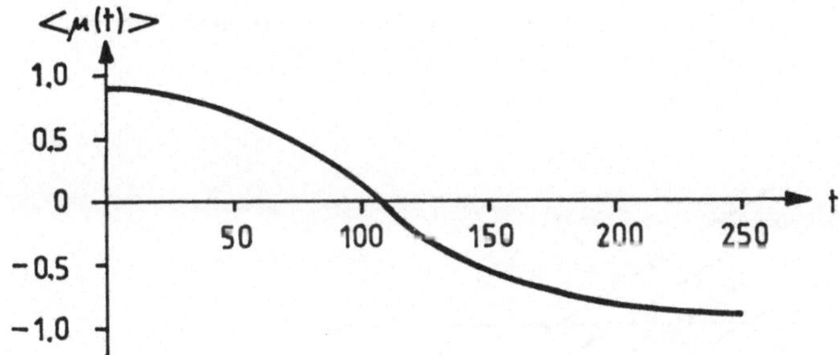

Fig. 1.13:
Order parameter $\langle\mu(t)\rangle$ (eq. (1.28)) versus time
(From: BINDER, 1977b)

been pointed out, this theory assumes that the equilibrium cluster distribution n_k is related to the formation energy F_k by:

$$n_k \sim \exp(- F_k/k_B T) \tag{1.31}$$

with

$$F_k/k_B T = - h\,k - \Gamma\,k^{2/3} \tag{1.32}$$

This equation is in agreement with eq. (1.3). The variable k gives the
number of reversed spins which form the cluster. The first term is the
contribution from the bulk free energy, i.e. the energy required to
flip k spins. h is the applied field corresponding to the supersaturation y. The second term is the surface free energy with Γ related to
the bulk surface tension σ associated with the interface between the
up and down spins by $\Gamma = (36\pi)^{1/3} \sigma/k_B T$ (HEERMANN et al., 1984). σ is
taken to be a constant, equal to the tension of a flat interf   and
independent of the quenching parameter (capillary approximation).
In eq. (1.32) the cluster are assumed to be compact and spherical in
contrast to a ramified structure, which is also discussed by HEERMANN
et al. (1984). These authors suppose further a model of modified droplets for an Ising system, where bonds between nearest-neighbour "up"
spins are diluted with a certain probability. This has the effect of
either reducing the size of the cluster or breaking up the Ising droplets into smaller ones.
Inserting eq. (1.32) into eq. (1.31) and taking the logarithm it holds

$$\ln n_k - hk = -\Gamma k^{2/3} + a \qquad (1.33)$$

where a comprises the proportionality factor in eq. (1.31), which is
classically taken to be the number density of monomers.
Fig. 1.14 presents the droplet size distribution. From eq. (1.33) in a
semilogarithmic plot a straight line should be observed, which is compared with data of the Monte Carlo simulations.

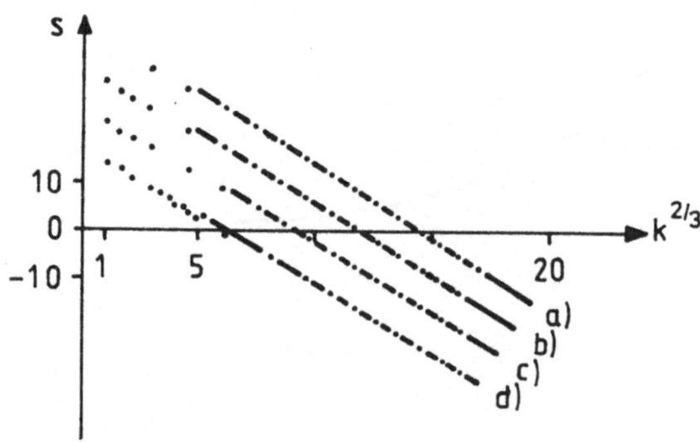

Fig. 1.14:
Variation of the surface part of the droplet formation energy S =
$\ln n_k$ - hk = $-\Gamma k^{2/3}$ +a (eq. (1.33)) with the droplet surface $k^{2/3}$ using
Ising droplets. The lines give the predictions of the theory, the dots
are the data from Monte Carlo simulations.
a) h = 0.55; b) h = 0.525; c) h = 0.485; d) h = 0.45

(From: HEERMANN et al., 1984)

It is to be seen, that the large clusters obey approximately the predictions of the classical nucleation theory. Only for small drops the
relation (eq. (1.31)) is not fulfilled, since the assumption of a compact cluster seems not to be valid. For such small clusters, consisting
of 10 or less monomers, KALOS et al. (1978) found empirical formulaes.
The parallel lines in Fig. 1.14 suggest that the surface tension is independent of the applied field h as also assumed in the classical nucleation theory, since h is chosen still close to the coexistence curve.
According to eqs. (1.17), (1.21) the classical nucleation rate can be
expressed by

$$I \sim n_k^{cr} \sim \exp\left\{ - 4\, \Gamma^3/27\, h^2 \right\} \tag{1.34}$$

n_k^{cr} denotes the number of clusters which overcome the nucleation barrier at the critical size k_{cr}, which classically is given by

$$k_{cr} = (2\,\Gamma/3\,h)^3 \tag{1.35}$$

Fig. 1.15 gives the critical droplet size; Fig. 1.16 presents the
steady state nucleation rate and the number of critical clusters, both
obtained from the theory and from simulations for a varying applied
field.
Further investigations on clustering and nucleation in a lattice gas
model are presented by Kalos, Penrose, Lebowitz and co-workers (KALOS
et al., 1978; PENROSE et al., 1978, 1983, 1984). As already mentioned,
they give empirical formulas to describe the behaviour of small clusters, and simulation data of the cluster distribution during the phase transition. Particularly, PENROSE and BUHAGIAR (1983) examine the
classical nucleation theory with microscopic considerations in a lattice gas model.
Finally we want to give two remarks. The first one is concerned with
the usually used assumption of an infinite system in classical nucleation theory. In Monte Carlo simulations periodic boundary conditions
and homogeneous exchange constants and fields are used in order to
have translational symmetry and thus avoid finite-size-effects as much
as possible. The presented results are expected to fail in the case of
a finite system. A second remark is due to the late stage of phase
transition, where processes like coagulation and competition of clusters (Ostwald ripening) seems to be important. This fact is not considered in the classical nucleation theory. With Monte Carlo methods
usually one cannot study the very late stage behaviour of a system due

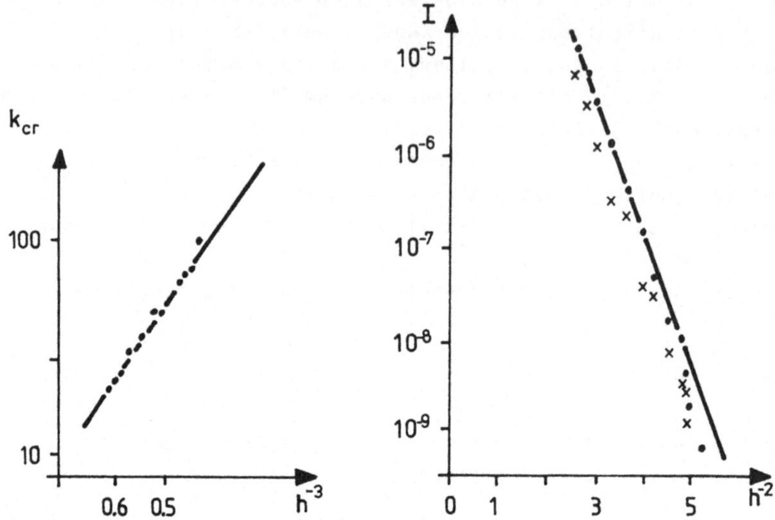

Fig. 1.15:
Critical cluster size k_{cr} obtained from the simulations (dots) and from the theory (eq. (1.35)) (solid line) versus $1/h^3$. $T/T_c = 0.59$

(From: HEERMANN et al., 1984)

Fig. 1.16:
Nucleation rate I plotted logarithmically versus $1/h^2$. The crosses give the number of critical clusters n_k, the dots the observed nucleation rate, as found by the simulations. The solid line is the prediction from the classical nucleation theory. $T/T_c = 0.59$

(From: HEERMANN et al., 1984)

to the required long computer time. Therefore most Monte Carlo simulations are carried out to study the dynamics of phase separation for times much before final thermodynamic equilibrium is reached.
Some ideas to describe the kinetics of the whole process of phase transition by nucleation in finite systems both in theory and computer simulations we suggest in the Chapters 4 - 6.

2. Thermodynamics of Heterogeneous Systems

2.1. Thermodynamic Premises of Classical Nucleation Theory

As it was already mentioned in chapter 1 the following premises of
thermodynamic nature are underlying classical nucleation theory, a
number of its modifications and generalizations; first, the existence
of a critical cluster size r_c; second, the possibility to express the
work of formation of such critical clusters by

$$W = \frac{1}{3} \sigma A_c \qquad A_c = 4 \pi r_c^2 \tag{2.1}$$

where σ is the surface tension for a planar interface between the
both phases; third, the possibility to describe the bulk contributions to the thermodynamic properties of the cluster by the macroscopic characteristics of the corresponding phases; fourth, that the probability P of formation of the critical clusters is proportional to

$$P \sim \exp(- \frac{W}{kT}) \tag{2.2}$$

The problems to investigate, now, are, whether the assumption, that
the surface tension of small clusters is equal to the corresponding
value for a planar interface (capillarity approximation) is valid,
under which conditions eq. (2.1) can be obtained and, in particular,
whether eq. (2.1) holds also for finite systems. To solve these problems an adequate description of surface contributions to the properties of the systems of interest is necessary.
There exists a number of different thermodynamic approaches to the
description of surface effects (GIBBS, 1878; VAN DER WAALS, 1893;
BAKKER, 1928; GUGGENHEIM , 1940, 1945, 1949; CAIIN, HILLTARD, 1958,
1959; ONO, KONDO, 1960; HILL, 1963; DEFAY et al., 1966; RUSANOV,
1967, 1978; GOODRICH, RUSANOV, 1981). These attempts can be divided
into two groups in dependence on whether the structure of the interfacial region is taken into account explicitly (e.g. van der Waals)
or not (e.g., Gibbs). To our opinion, the Gibbs theory of heterogeneous systems is both in its basic axioms and in the logical development of the theory not only historically the first but at the same
time the most organic extension of classical thermodynamics to the
description of surface effects. This theory is always applicable, if
not the structure of the interfacial region itself is the object of
investigation.
Since in our analysis of formation and growth of clusters of a new
phase a detailed knowledge of the interfacial region is, in general,
not needed, we will describe surface properties based on Gibbs'

approach. First, therefore, the main postulates and some results of
Gibbs' theory are summarized to be extended then to allow also a
description of non-equilibrium states. This extension of Gibbs' theory is more general compared with the theory of DEFAY et al. (1966),
in which from the very beginning mechanical and thermal equilibrium
between the different phases is assumed.

2.2. Gibbs' Theory of Heterogeneous Systems

The coexistence of different phases implies the existence of interfacial regions, where the intensive thermodynamic quantities, e.g.,
the densities of the different components, vary rapidly (see, e.g.
RUSANOV, 1978; Faraday Symposia, 1981). The whole system can be considered, therefore, as consisting of two nearly homogeneous regions
(for which classical thermodynamics is applicable) and an inhomogeneous interfacial region (see Fig. 2.1).
The first postulate of Gibbs' theory (GIBBS, 1878) is that for fixed
positions of the surfaces X_1, X_2, dividing the homogeneous from the
inhomogeneous parts of the system, the inhomogeneous region can
be described thermodynamically in the same way as homogeneous media.
According to classical thermodynamics it is possible to write then

$$dU = TdS + \sum_{i=1}^{k} \mu_i dn_i \tag{2.3}$$

U, S, T, μ_i and n_i are the inner energy, the entropy, the absolute
temperature, the chemical potential and the number of moles of the
i-th component (i=1,2,...,k) in the inhomogeneous region II. From the
general thermodynamic equilibrium condition (2.4)

$$(\delta S)_{U,V,n_i} \leqslant 0 \tag{2.4}$$

Gibbs concluded, that in an equilibrium state the relations

$$T_I = T_{II} = T_{III} = T \qquad \mu_{iI} = \mu_{iII} = \mu_{iIII} = \mu_i \tag{2.5}$$

have to be fulfilled, the subscripts I, II, III denoting the values
of the thermodynamic parameters in the different regions. δS in eq.
(2.4) is the change of the entropy due to possible variations of the
state of the system not contradicting the constraints U = const., V =
const., n_i = const., V being the volume. These equilibrium conditions
are assumed to be fulfilled in Gibbs' theory.

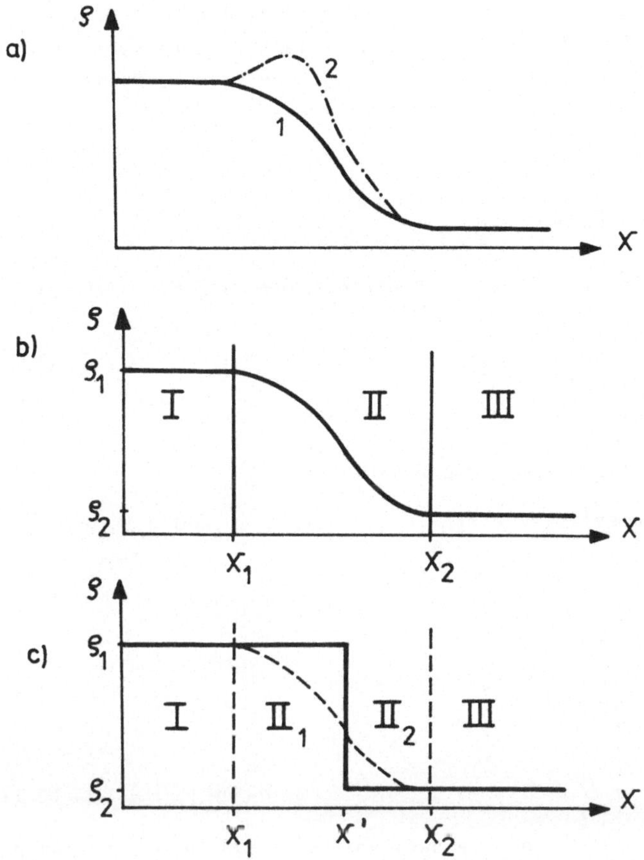

Fig. 2.1:
a) Density profile $\varrho(X)$ for a coexistence of two phases with (curve 2) and without (curve 1) surface active substances.
b) Possible dependence of the density ϱ as a function of a coordinate X for a two-phase system. The X-axis is directed perpendicular to the interfacial region II, which lies between the homogeneous parts I and III.
c) The Gibbs model, where the real distribution is replaced by two homogeneous parts (I+II$_1$, II$_2$+III), divided by a dividing surface at the position X'.

(From: MAHNKE, SCHMELZER, THIESSENHUSEN, 1985)

Further, in Gibbs' theory the inhomogeneous region II is considered
as consisting of two homogeneous parts, marked by the subscripts 1
and 2, respectively, with the same properties as the neighbouring
phases I and III, divided by a dividing surface X' (see Fig. 2.1). For
both these homogeneous parts an equation of the form (2.3) is valid,
again, if the dividing surface X' is fixed also.

$$dU_1 = T_1 dS_1 + \sum_i \mu_{i1} dn_{i1} \tag{2.6}$$

$$dU_2 = T_2 dS_2 + \sum_i \mu_{i1} dn_{i2}$$

From the equilibrium conditions immediately the eqs. (2.7) are obtained.

$$T_1 = T_2 = T \qquad \mu_{i1} = \mu_{i2} = \mu_i \tag{2.7}$$

By substracting eq. (2.6) from eq. (2.3) we get

$$d(U-U_1-U_2) = Td(S-S_1-S_2) + \sum_i \mu_i d(n_i - n_{i1} - n_{i2}) \tag{2.8}$$

Introducing the superficial energy U_0 , entropy S_0 and mole number
n_{io} by

$$U_0 = U-U_1-U_2 \qquad S_0 = S-S_1-S_2 \qquad n_{io} = n_i - n_{i1} - n_{i2} \tag{2.9}$$

we get

$$dU_0 = TdS_0 + \sum_i \mu_i dn_{io} \tag{2.10}$$

and

$$U = U_1 + U_2 + U_0 \qquad S = S_1 + S_2 + S_0 \qquad n_i = n_{i1} + n_{i2} + n_{io} \tag{2.11}$$

The quantities U_0, S_0 and n_{io} (excess values) describe the deviations
from the additivity, in Gibbs' approach they can be considered formally as thermodynamic parameters of the dividing surface.
The relations (2.10) and (2.11) are independent of the position of
the surfaces X_1 and X_2, if these are placed inside the homogeneous
parts. So, X_1 and X_2 can be identified with the external boundaries
of the system.
The relations (2.10) and (2.11) are also independent of the choice of
the position of the dividing surface X', though the values of U_0, S_0,
n_{io} depend, of course, on it.
Variations of the position and form of the interfacial region in
Gibbs' theory are described by changes of the parameters specifying
the properties of the dividing surface. Since a sufficiently small
surface element is described completely by its surface area dA and the

two principal curvatures c_1 and c_2 or principal radii r_1 and r_2, in accordance with the basic ideas of classical thermodynamics, Gibbs' postulated (3rd axiom), that, if changes of the form and position of the dividing surface are taken into account, dU_0 has to be written in the following way:

$$dU_0 = TdS_0 + \sum_i \mu_i dn_{io} + \sigma\, dA + C_1 dc_1 + C_2 dc_2 \qquad (2.12)$$

σ, C_1 and C_2 are functions of the independent variables describing the state of the system.

Eq. (2.12) is strictly valid for isotropic media only, but in a good approximation it can be used also for a description of crystallites, if not the effects due to the anisotropy are the object of investigation (SKRIPOV, KOVERDA, 1984; WOODRUFF, 1973; JACKSON, 1984). Such effects are not considered here and so we can assume a spherical shape of the clusters.

For spherical clusters the equations $c_1 = c_2 = c$ and $C_1 = C_2 = C/2$ hold and eq. (2.12) reads

$$dU_0 = TdS_0 + \sum_i \mu_i dn_{io} + \sigma\, dA + Cdc \qquad (2.13)$$

Gibbs showed further, that a special choice of the dividing surface is possible for which the parameter C becomes equal to zero. This special dividing surface was denoted by Gibbs as surface of tension, the corresponding value of σ as surface tension.

Choosing the surface of tension as the dividing surface eq. (2.13) can be reduced to

$$dU_n^{(s)} = TdS_0^{(s)} + \sum_i \mu_i dn_{io}^{(s)} + \sigma^{(s)} dA^{(s)} \qquad (2.14)$$

Integration of eq. (2.13) over a finite area with constant values of T, μ_i, σ results in

$$U_0 = TS_0 + \sum_i \mu_i n_{io} + \sigma A \qquad (2.15)$$

A derivation of eq. (2.15) and a comparison with eq. (2.13) leads to the general form of the Gibbs adsorption equation (BUFF, 1951)

$$S_0 dT + Ad\sigma + \sum_i n_{io} d\mu_i = C\, dc \qquad (2.16)$$

Choosing the surface of tension as the dividing surface eq. (2.16) is reduced to

$$S_0^{(s)} dT + A^{(s)} d\sigma^{(s)} + \sum_i n_{io}^{(s)} d\mu_i = 0 \qquad (2.17)$$

Taking into account a possible variation of the volume of the system eqs. (2.6) have to be written as

$$dU_1 = T_1 dS_1 - p_1 dV_1 + \sum_i \mu_{i1} dn_{i1}$$

$$dU_2 = T_2 dS_2 - p_2 dV_2 - \sum_i \mu_{i2} dn_{i2}$$

$$(2.18)$$

Moreover, U_1 and U_2 can be expressed by

$$U_1 = T_1 S_1 - p_1 V_1 + \sum_i \mu_{i1} n_{i1}$$

$$U_2 = T_2 S_2 - p_2 V_2 + \sum_i \mu_{i2} n_{i2}$$

$$(2.19)$$

From the equilibrium conditions (2.7) and eqs. (2.11) - (2.12) we get further

$$dU = TdS - p_1 dV_1 - p_2 dV_2 + \sum_i \mu_i dn_i + \sigma dA + Cdc \tag{2.20}$$

Since the equilibrium condition (2.4) is equivalent to

$$(\delta U)_{S,V,n_i} > 0 \tag{2.21}$$

the condition for a mechanical equilibrium reads

$$p_1 - p_2 = \sigma \frac{dA}{dV_1} + C \frac{dc}{dV_1} \tag{2.22}$$

If the surface of tension is chosen as the dividing surface (C=0) and a spherical form of the clusters is assumed, then the Young-Laplace equation (2.23) is easily obtained from eq. (2.22);

$$p_1 - p_2 = 2 \sigma^{(s)} / r_s \tag{2.23}$$

Moreover, if the position of the dividing surface is changed the physical state of the system being unvaried ($dT = d\mu_i = 0$), then from eq. (2.16) we obtain

$$A[d\sigma] = C [dc] \tag{2.24}$$

the brackets indicating the special variation (redefinition of the dividing surface) considered.
In the same way we get from eq. (2.20)

$$p_1 - p_2 = \frac{dA}{dV_1} + C \left[\frac{dc}{dV_1} \right] \tag{2.25}$$

and combining eqs. (2.24) and (2.25) a generalized Young-Laplace equation results.

$$p_1 - p_2 = \frac{2\sigma}{r} + \left[\frac{d\sigma}{dr} \right] \tag{2.26}$$

This equation describes variations of the value of the surface tension due to a change of the position of the dividing surface the physical state of the system being unchanged (p_1 and p_2, e.g., are constant).
It follows that the definition of the dividing surface proposed by ONO and KONDO (1960)

$$\left[\frac{d\sigma}{dr}\right] = 0 \qquad (2.27)$$

is equivalent to Gibbs' definition $C = 0$. Moreover, from the extremum condition it can be obtained easily that the surface of tension corresponds to a minimum possible value of the surface tension for a fixed physical state of the system (SCHMELZER, 1985a; see also RUSANOV, 1967, 1978).
An equation of the same form as eq. (2.26) but with a very different meaning can be derived for isothermal processes in one-component systems if the equimolecular dividing surface is chosen as the dividing surface. The position of this special dividing surface is determined by $n_0 = 0$. So we get

$$A d\sigma = C dc \qquad (2.28)$$

again and from eq. (2.22) it follows (see also BUFF, 1951; ROWLINSON and WIDOM, 1982)

$$p_1 - p_2 = \frac{2\sigma^{(e)}}{r_e} + \frac{d\sigma^{(e)}}{dr_e} \qquad (2.29)$$

$\sigma^{(e)}$ is the value of the surface tension referred to the equimolecular dividing surface with the radius r_e.
Based on these general results we will go over now to the analysis of the problem of the curvature dependence of surface tension of droplets and bubbles.

2.3. Curvature Dependence of Surface Tension

Already by GIBBS (1878), W. THOMSON (1906, 1929) and BAKKER (1928) the problem was discussed, down to which size of a drop the macroscopic value of the surface tension can be used for a description of its thermodynamic properties, if and in which form the surface tension depends on the size of the droplets.
Gibbs came to the conclusion, that curvature corrections to σ are of importance for small drops only comparable in size with the width of the inhomogeneous region between both phases and that the surface ten-

sion decreases with decreasing droplet size. Bakker expressed the
opinion that the surface tension is independent of the size of the
drops while Thomson assumed a non-monotonic behaviour of σ as a
function of the droplet radius.

The problem of a curvature dependence of surface tension became important, in particular, with the development of nucleation theory, since,
according to eq. (2.1), the nucleation rate depends significantly on
the value of the surface tension. Based on the work of GIBBS (1878),
GUGGENHEIM (1940), TOLMAN (1948, 1949), KIRKWOOD and BUFF (1949) the
possible curvature dependence of surface tension was widely discussed
in the literature (see, e.g., ONO, KONDO, 1960 and cited there refe-
rences; SCHERBAKOW et al., 1961; DERJAGUIN, PROKHOROV, 1974;
BELOSLUDOV, NABUTOVSKI, 1975; RUSANOV, BRODSKAYA, 1977; RASMUSSEN,
1982; RASMUSSEN et al., 1983; FISHER, ISRAELACHVILI, 1980; FALLS
et al., 1981, 1983; BAIDAKOV, SKRIPOV, 1982; HEMINGWAY et al., 1981;
PHILLIPS, 1984; VOGELSBERGER, MARX, 1976; VOGELSBERGER et al., 1983,
1985; NONNENMACHER, 1976, 1977; KOTAKE, GLASS, 1978; RUSANOV, 1967,
1978; THOMPSON et al., 1984; WINGRAVE et al., 1981; GUERMEUR et al.,
1985). In some investigations of this problem it is explicitely or
implicitely expressed, that the obtained results are of purely thermodynamic and, therefore, of a very general nature.

However, already for a planar interface the surface tension cannot be
calculated by thermodynamic methods only and, consequently, we cannot await, that a calculation of the curvature dependence of surface
tension is possible based only on thermodynamics. So it is studied
here, which results can be obtained from purely thermodynamic investigations and which results are the consequence of additional assumptions, approximations involved in the calculations.

The equations determining the curvature dependence of surface tension
of clusters in equilibrium with the surrounding medium for one-component systems at a constant temperature are given by (see eqs. (2.16),
(2.22); SCHMELZER, 1985b):

$$Ad\sigma + n_0 d\mu = Cdc$$

$$p_\alpha - p_\beta = 2\sigma/r_\alpha + C(dc/dV_\alpha) \qquad \mu_\alpha = \mu_\beta = \mu \tag{2.30}$$

If, as it is usually done, the surface of tension is chosen as the
dividing surface (C=0), we get instead of eq. (2.30)

$$d\sigma^{(s)} = -\Gamma_0 d\mu \qquad \Gamma_0 = n_0^{(s)}/A^{(s)}$$

$$p_\alpha - p_\beta = 2\sigma^{(s)}/r_s \qquad \mu_\alpha = \mu_\beta = \mu \tag{2.31}$$

The subscript α specifies here and in the following the thermodynamic parameters of the clusters (e.g., drops or bubbles), β the parameters of the surrounding the clusters medium.

It is of importance to note here, that a definite choice of the dividing surface is necessary, otherwise the problem of determination of the curvature dependence is not properly defined (for a critical discussion of some attempts see SCHMELZER, MAHNKE, 1986; SCHMELZER, 1985a).

Since the chemical potential ($\mu = \mu_\alpha$) can be considered as a function of the molar density of the clusters ρ_α and temperature T, the surface tension is also a function of ρ_α and T. There exists, consequently, only an implicite dependence of σ on the curvature due to the dependence of ρ_α on the size of the cluster. It follows, that the assumptions of a curvature dependence of surface tension and the assumption of an incompressible cluster phase are incompatible from a theoretical point of view. On the other hand, it follows, that the validity of the capillarity approximation can be interpreted thermodynamically as a consequence of the incompressibility of the cluster.

From the eq. (2.31) the following differential equation for $\sigma^{(s)}$ can be obtained (GIBBS, 1878; TOLMAN, 1949):

$$\frac{d\sigma^{(s)}}{\sigma^{(s)}} = -\frac{2\delta}{1 + \frac{2\delta}{r_s}}\, d\left(\frac{1}{r_s}\right) \tag{2.32}$$

The Tolman parameter δ , which depends, in general, also on the size of the cluster, is defined by

$$\delta = \frac{\Gamma_e}{\rho_\alpha - \rho_\beta} \qquad \delta_\infty - \lim_{r_s \to \infty} \delta \tag{2.33}$$

Integration of eq.(2.32) results in (MAHNKE, SCHMELZER, 1985):

$$\sigma^{(s)} = \frac{\sigma_\infty}{1 + \frac{2\delta(r_s)}{r_s}}\, \exp\left\{2\int_\infty^{r_s} \frac{\frac{1}{r}\frac{\partial\delta}{\partial r}}{1 + \frac{2\delta(r)}{r}}\, dr\right\} \tag{2.34}$$

σ_∞ and δ_∞ are the values of the surface tension and the Tolman coefficient for a planar interface between the different phases.

The eqs. (2.32) and (2.34) are general thermodynamic relations; they show, that the curvature dependence of surface tension is determined by the dependence of the Tolman coefficient on r_s. The functions $\delta = \delta(r_s)$ and, consequently, $\sigma^{(s)} = \sigma^{(s)}(r_s)$ cannot be calculated by thermodynamic methods. Despite this fact, eqs. (2.33) and (2.34) are useful relations, since relatively simple assumptions concerning δ lead to relatively complicated expressions for $\sigma^{(s)}$.

Gibbs, e.g., proposed

$$\delta \ll r_s \qquad\qquad \delta = \delta_\infty > 0 \tag{2.35}$$

which results in

$$\sigma^{(s)}(r_s) = \sigma_\infty \exp\left\{ - \frac{2\,\delta_\infty}{r_s} \right\} \tag{2.36}$$

As one approximation Tolman assumed $\delta = \delta_\infty > 0$ and obtained

$$\sigma^{(s)}(r_s) = \frac{\sigma_\infty}{1 + \dfrac{2\,\delta_\infty}{r_s}} \tag{2.37}$$

The Tolman coefficient is nearly equal (in the limit $r_s \to \infty$ equal) to the distance between two dividing surfaces, the equimolecular dividing surface and the surface of tension (SCHMELZER, 1985a,b).

Based on this interpretation we proposed the following ansatz for δ (SCHMELZER, MAHNKE, 1986; MAHNKE, SCHMELZER, 1985)

$$\delta = \delta_\infty \frac{r_s}{r_s + a} \tag{2.38}$$

and

$$\delta = \delta_\infty \frac{r_s + b}{r_s + a} \qquad\qquad b < 0 \tag{2.39}$$

a, b and δ_∞ are parameters describing the system considered, they depend on temperature but not on r_s.

If $\Gamma_0 > 0$ ($\Gamma_0 < 0$) then the eq. (2.38) describe the curvature dependence of δ and $\sigma^{(s)}$ for drops (bubbles), eq. (2.39) describes bubbles (drops). The relations (2.38) - (2.39) were motivated by the requirement, that for large values of r_s δ has to become independent of the size of the cluster and that physically reasonable (non-divergent) solutions for σ have to be obtained (MAHNKE, SCHMELZER, 1985; SCHMELZER, MAHNKE, 1986). As a special consequence it follows, that the parameter b has to be less than zero.

It is obvious, that in the case a < 0 physically reasonable values for $\sigma^{(s)}$ are obtained only for $r_s > |a|$ since for $r_s \to |a|$ δ tends to infinity and the cluster model becomes inappropriate (see eq. (2.32)). The results obtained after a substitution of eq. (2.38) into (2.34) are

$$\sigma^{(s)} = \sigma \left\{ \frac{r_s}{a + 2\delta_\infty + r_s} \right\}^{\frac{2\delta_\infty}{a+2\delta_\infty}} \qquad a+2\delta_\infty \neq 0 \qquad\qquad (2.40)$$

In the case $a = -2\delta_\infty$ Gibbs' equation (2.36) results.
The curves for $\sigma^{(s)}$ are presented in Fig. 2.2 for different values
of the parameter a. First, it can be noted, that a number of different formulas for the curvature dependence of surface tension of drops
($\Gamma_0 > 0$) or bubbles ($\Gamma_0 < 0$) derived by different authors earlier
are included in eq. (2.40) as special cases (e.g., a=0, Tolman; a=
$-2\delta_\infty$, Gibbs; a=$-3\delta_\infty$, RASMUSSEN (1982)). Moreover, a limit of appli-
cability of these different equations is established ($r_s > |a|$).
Further it can be mentioned, that for a very different behaviour of
the function $\delta = \delta(r_s)$ the functions $\sigma^{(s)} = \sigma^{(s)}(r_s)$ are quite similar, indicating, that $\sigma^{(s)}$ is mainly determined by the sign of δ.

The results obtained for the case described by eq. (2.29) can be
approximated by the equation (2.41) (for the exact solution see
SCHMELZER, MAHNKE, 1986):

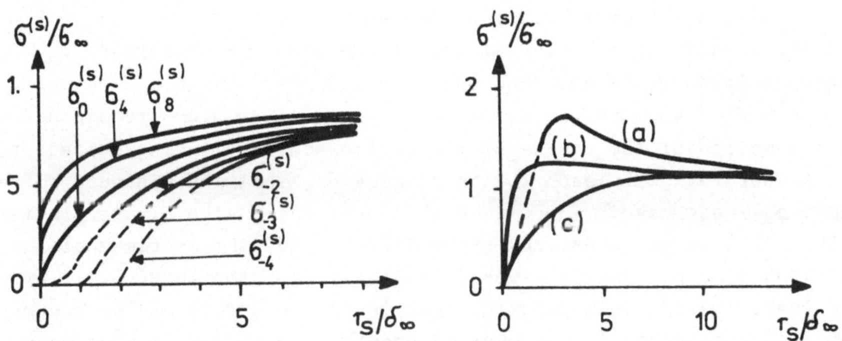

Fig. 2.2 (left), Fig. 2.3 (right):
Curvature dependence of surface tension as described by eqs. (2.38),
(2.40) and (2.39) and (2.41), respectively. For $\Gamma_0 > 0$ ($\Gamma_0 < 0$) the behaviour as presented by Fig. 2.2 corresponds to drops (bubbles), Fig.
2.3 shows the curvature dependence of surface tension of bubbles
(drops). By a dashed curve the interval is indicated, where according
to our derivation no solution exists ($r_s < |a|$ for a<0). Since the
parameters are functions of the temperature, this non-existence of solutions of eq. (2.34) means, that for the given values of the temperature there cannot be formed a cluster in equilibrium with the surrounding medium.
(From: SCHMELZER, MAHNKE, 1986)

$$\sigma^{(s)}(r_s) = \frac{r_s \sigma_\infty}{\sqrt{r_s^2 + Ar_s + B}} \qquad A,B - \text{constants} \qquad (2.41)$$

in agreement with e.g. BELOSLUDOV, NABUTOVSKI (1975); BAIDAKOV, SKRIPOV (1982); FISHER, ISRAELACHVILI (1980) and GUERMEUR et al. (1985). The behaviour of the solutions represented by eq. (2.41) are shown in Fig. 2.3.

It was already mentioned, that our formulas are in agreement with the (up to now) limited number of statistical mechanical approaches to the problem of curvature dependence of surface tension (KIRKWOOD, BUFF, 1949; FALLS et al., 1981, 1983; ROWLINSON, WIDOM, 1982; HENDERSON, ROWLINSON, 1984; GUERMEUR et al., 1985). Most of approaches are valid only for the special type of the interaction potential assumed and contain, in addition, sometimes serious approximations. So it is of interest to find out exact results for the curvature dependence of surface tension, at least, for a particular system.

Such an analysis was carried out (SCHMELZER, 1986a) based on molecular dynamics simulations of drops with a Lennard-Jones interaction potential reported by THOMPSON et al. (1984) (see also RUSANOV, BRODSKAYA, 1977). Our analysis was based only upon exact thermodynamic relations, in particular, on the eqs. (2.26) and (2.29). In Fig. 2.4 the results of the simulations of the pressure difference $p_\alpha - p_\beta$ the calculated by us and THOMPSON et al. (1984) values of the surface tension $\sigma^{(s)}$ and the radius of the surface of tension are expressed as a function of the radius of the equimolecular dividing surface r_e. Since thermodynamics leads to the conclusion, that the pressure difference $p_\alpha - p_\beta$ has to increase with a decreasing size of the cluster (SCHMELZER, 1986a; SCHMELZER, SCHWEITZER, 1985) the extremum of the function $p_\alpha - p_\beta = f(r_e)$ gives an estimation of the lower limit of applicability of thermodynamics. For the given values of the parameters it corresponds to a number of about 170 particles in the drop.

Finally, we would like to show, how the thermodynamic investigations outlined here for one component can be generalized to multicomponent systems. From the very beginning the surface of tension is chosen as the dividing surface, which leads to the following equations for the determination of the curvature dependence of surface tension in multicomponent systems (T is constant, again):

$$A^{(s)}d\sigma^{(s)} + \sum_i n_{io}^{(s)}d\mu_i = 0 \qquad p_\alpha - p_\beta = 2\sigma^{(s)}/r_s$$

$$\mu_i = \mu_{i\alpha} = \mu_{i\beta} \qquad i = 1,2,\ldots,k \qquad (2.42)$$

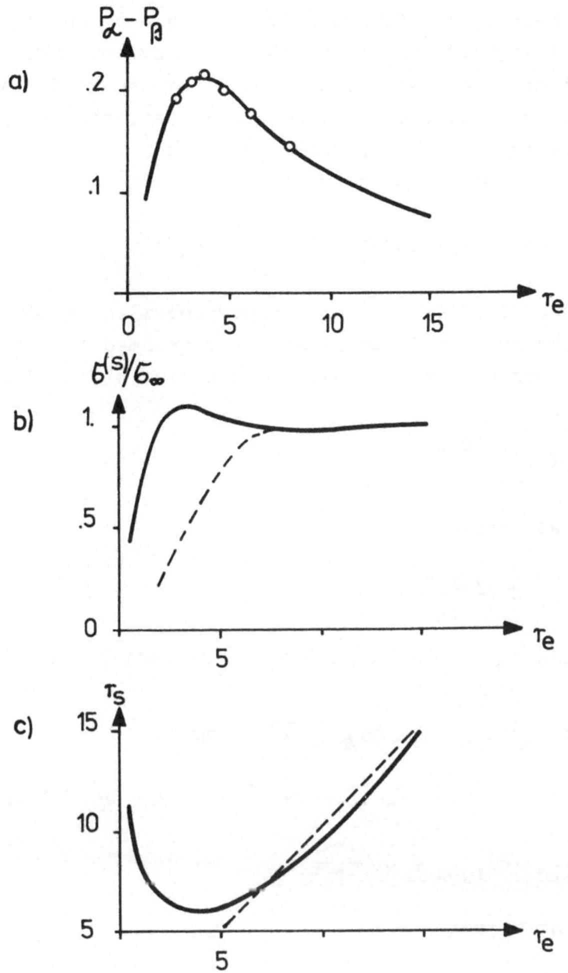

Fig. 2.4:
a) Interpolation of the pressure difference as a function of the
   radius of the equimolecular dividing surface r_e.
b) Values of the surface tension $\sigma^{(s)}$ as a function of r_e. By a dashed
   curve the results as calculated by THOMPSON et al. (1984) and
   RUSANOV, BRODSKAYA (1977) are shown.
c) The radius of the surface of tension as a function of r_e obtained
   via the general thermodynamic route. The dashed line corresponds
   to $r_s = r_e$.
(From: SCHMELZER, 1986a)

In this general case it is not possible to obtain an analogous to
eq. (2.32) result without specifying the thermodynamic properties of
the system considered. Let us assume here, that one of the phases can
be considered as an ideal gas (it means we investigate either the formation of bubbles in a liquid or drops in a gas). The chemical potential of the i-th component can be written then as

$$\mu_{igas} = \mu_i(T,p) + R_g T \ln \frac{p_{gas} x_i}{p} \tag{2.43}$$

x_i being the molar fraction of the i-th component in the gas, R_g the
universal gas constant. If in addition it is assumed that the molar
fractions do not depend significantly on the size of the cluster, we
get

$$d\mu_{igas} = R_g T \, d(\ln \frac{p_{gas}}{p}) \tag{2.44}$$

'From eq. (2.42) we obtain

$$dp_\alpha - dp_\beta = d \left(\frac{2\sigma^{(s)}}{r_s}\right) \tag{2.45}$$

Expressing p_α and p_β by the Gibbs-Duhem equations for the bulk
phases

$$dp_\alpha = \sum_i \rho_{i\alpha} \, d\mu_{i\alpha} \qquad dp_\beta = \sum_i \rho_{i\beta} \, d\mu_{i\beta} \tag{2.46}$$

and taking into account the equilibrium conditions eq. (2.47) results;

$$\sum_i (\rho_{i\alpha} - \rho_{i\beta}) \, d\mu_{igas} = d\left(\frac{2\sigma^{(s)}}{r_s}\right) \tag{2.47}$$

Introducing the notations

$$\Gamma_{io} = \frac{n_{io}^{(s)}}{A^{(s)}} \tag{2.48}$$

the Gibbs adsorption isotherm reads

$$d\sigma^{(s)} = - \sum_i \Gamma_{io} \, d\mu_{igas} \tag{2.49}$$

The variations of the chemical potentials can be eliminated by eq.
(2.44) and again a differential equation of the type (2.32) is obtained

$$\frac{d\sigma^{(s)}}{\sigma^{(s)}} = - \frac{2\delta_k}{1 + \frac{2\delta_k}{r_s}} \, d(\frac{1}{r_s}) \tag{2.50}$$

where δ_k is defined, now, by

$$\delta_k = \frac{\sum \Gamma_{io}}{\sum (\rho_{i\alpha} - \rho_{i\beta})} \tag{2.51}$$

Note, that δ_k for a given value of the sum $\sum \Gamma_{io}$ has different signs in dependence whether the formation of drops or bubbles is considered.

The same considerations as carried out for the one-component can be repeated now for the multicomponent case. So in dependence on the actual value of the generalized Tolman parameter δ_k we get a behaviour as presented in Fig. 2.2 or 2.3, respectively. The curves can become more complicated, if changes of the molar fraction as a function of the cluster size are taken into account. These deviations should be small, if the surface tension does not depend significantly on the composition. It means, that the absolute value of the first term in eq. (2.52) has to be small compared with the second one.

$$d\sigma^{(s)} = - R_g T \sum \Gamma_{io} \, d(\ln x_i) - R_g T \, d(\ln \frac{p_{gas}}{p}) \tag{2.52}$$

In the subsequent discussions always the surface of tension is chosen as the dividing surface. Moreover, since we are interested mainly in a qualitative insight into the mechanism of first-order phase transitions the capillarity approximation is used ($\sigma^{(s)} = \sigma_\infty$). This approximation is always valid for the stage of growth of sufficiently large clusters. In the chapter 5.5. it will be shown, moreover, that the possible curvature dependence of surface tension does not lead to a qualitatively different behaviour compared with the results obtained based on this approximation.

2.4. Heterogeneous Systems in Non-Equilibrium States and the Principle of Inner Equilibrium

Heterogeneous systems consisting of clusters of a new phase in the otherwise homogeneous medium represent, in general, non-equilibrium states. Therefore, it is necessary to discuss the way of a thermodynamic description of such states. The thermodynamic description used here is based on the principle of inner equilibrium, which is discussed in detail (SCHMELZER, 1984, 1985a,c; see also RUSANOV, 1967, 1978) in the following.

In classical thermodynamics of homogeneous systems variations of the state of the system are considered, which proceed via a sequence of

equilibrium states. The changes of the thermodynamic functions according to the first and second law of thermodynamics for such processes
can be expressed by

$$dU = TdS - pdV + \sum_i \mu_i dn_i \qquad (2.53)$$

where U, the inner energy, is given by

$$U = TS - pV + \sum_i n_i \mu_i \qquad (2.54)$$

A derivation of eq. (2.54) and comparison with eq. (2.53) leads to the
Gibbs-Duhem equations (2.55) which are valid also only for reversible
processes;

$$SdT - Vdp + \sum_i n_i d\mu_i = 0 \qquad (2.55)$$

The necessary conditions for a thermodynamic equilibrium state can be
formulated as

$$\delta\Phi = 0 \qquad (2.56)$$

where Φ is the appropriate thermodynamic potential for the given thermodynamic constraints.
The state, determined by eq. (2.56), is a stable equilibrium state if
in addition eq. (2.57) is fulfilled (KUBO, 1968):

$$\delta S\,\delta T - \delta p\,\delta V + \sum_i \delta\mu_i\,\delta n_i > 0 \qquad (2.57)$$

The assumption of an inner equilibrium is equivalent to the demand,
that the relations (2.53) - (2.55) and (2.57) are fulfilled for the
bulk phases and the analogous relations (2.12) - (2.17) hold for the
surface contributions. It is supposed, therefore, that the bulk and
surface phases are both in a state near the thermodynamic equilibrium
but no equilibrium between the different phases is, in general, assumed. The principle of an inner equilibrium is, therefore, quite similar to the principle of local equilibrium used in thermodynamics of
non-equilibrium processes (HAASE, 1963; GUROV, 1978). It leads to a
number of consequences, which are discussed in detail elsewhere
(KUBO, 1968; SCHMELZER, 1985a,c).
Based on the principle of an inner equilibrium now the thermodynamic
description of the simplest case - one cluster in the otherwise homogeneous medium - is developed. We postulate (postulate 1), that the
relations (2.11) hold also in non-equilibrium states. These equations
we write, now, as follows;

$$U = U_\alpha + U_\beta + U_0 \qquad (2.58)$$

where according to eqs. (2.15) and (2.54) U_α, U_β and U_0 can be written as

$$U_\alpha = T_\alpha S_\alpha - p_\alpha V_\alpha + \sum_i \mu_{i\alpha} n_{i\alpha}$$

$$U_\beta = T_\beta S_\beta - p_\beta V_\beta + \sum_i \mu_{i\beta} n_{i\beta} \qquad (2.59)$$

$$U_o = T_o S_o + \sigma A + \sum_i \mu_{io} n_{io}$$

α specifies the parameters of the cluster and β of the medium, respectively. The variations of U according to eq. (2.58) can be expressed by

$$dU = dU_\alpha + dU_\beta + dU_o \qquad (2.60)$$

or by (see eq. (2.14) and (2.53))

$$dU = T_\alpha \, dS_\alpha - p_\alpha \, dV_\alpha + \sum \mu_{i\alpha} \, dn_{i\alpha} + T_\beta \, dS_\beta - p_\beta \, dV_\beta +$$
$$+ \quad \mu_{i\beta} \, dn_{i\beta} + T_o dS_o + \sigma dA + \sum \mu_{io} dn_{io} \qquad (2.61)$$

The problem arises now, how to determine μ_{io} and T_o. In agreement with
DEFAY et al. (1966), PRIGOGINE and BELLEMANS (1980, "a surface phase
has no real autonomy, in general"), ROWLINSON and WIDOM (1982 "we
cannot measure or define unambiguously and independently the thermodynamic properties of the surface phase", p. 33) we demand, that the
intensive variables T_o and μ_{io} are equal to the corresponding quantities of the phase with the higher density (postulate 2). If, e.g.,
the development of clusters with a higher compared with the surrounding medium density is considered (e.g., drops in the gas), then the
eqs. (2.59) - (2.61) read:

$$U - T_\alpha \tilde{S}_\alpha - p_\alpha V_\alpha + \sum_i \mu_{i\alpha} \tilde{n}_{i\alpha} + T_\beta S_\beta - p_\beta V_\beta + \sum_i \mu_{i\beta} n_{i\beta} + \sigma A$$

$$dU = T_\alpha \, d\tilde{S}_\alpha - p_\alpha \, dV_\alpha + \sum_i \mu_{i\alpha} \, d\tilde{n}_{i\alpha} + T_\beta \, dS_\beta - p_\beta \, dV_\beta +$$
$$+ \sum_i \mu_{i\beta} \, dn_{i\beta} + \sigma \, dA \qquad (2.62)$$

Here the notations (2.63) are used.

$$\tilde{S}_\alpha = S_\alpha + S_o \qquad\qquad \tilde{n}_i = n_{i\alpha} + n_{io} \qquad (2.63)$$

Considering as the constraints

$$S = \tilde{S}_\alpha + S_\beta = \text{const.} \qquad n_i = \tilde{n}_{i\alpha} + n_{i\beta} = \text{const.}$$
$$V = V_\alpha + V_\beta = \text{const.} \qquad\qquad\qquad (2.64)$$

the necessary equilibrium conditions (2.56) are given by

$$\delta U = (T_\alpha - T_\beta)\, \delta \tilde{S}_\alpha + (-p_\alpha + p_\beta + \sigma\, dA/dV_\alpha)\, \delta V_\alpha +$$
$$+ \sum_i (\mu_{i\alpha} - \mu_{i\beta})\, \delta \tilde{n}_{i\alpha} = 0 \tag{2.65}$$

leading to

$$T_\alpha = T_\beta \qquad p_\alpha - p_\beta = 2\sigma/r \qquad \mu_{i\alpha} = \mu_{i\beta} \tag{2.66}$$

In the same way, other thermodynamic potentials can be introduced. According to postulate 1 the potentials can be expressed as

$$\Phi = \Phi_\alpha + \Phi_\beta + \Phi_o \tag{2.67}$$

In table 2.1 the equations determining the thermodynamic potentials are summarized for different constraints. For heterogeneous systems there exist, in general, different possibilities of definition of such potentials (see also MUTAFTSCHIEV, 1982; RUSANOV, 1978; YANG, 1983). The definitions used here have the advantage, that there exists a direct relation to the work of formation of critical clusters. In general, the difference between the thermodynamic potentials describing the heterogeneous system - cluster in the medium - and the homogeneous initial state for fixed values of the independent variables equals to the work of formation of the cluster (LANDAU, LIFSHITZ, 1978; RUSANOV, 1967, 1978);

$$W = \Phi_{het} - \Phi_{hom} \tag{2.68}$$

To avoid some unnecessary complications, moreover, for the definition of F and G the temperature T is assumed here to be constant.
The postulate 2 leads to the conclusion, that the surface tension of a cluster not in equilibrium with the surrounding medium is determined by the intensive variables of the phase with the higher density. If, e.g., the cluster has a higher compared with the surrounding medium density then according to postulate 2 the Gibbs adsorption equation reads

$$S_o dT + A d\sigma + \sum_i n_{io} d\mu_{i\alpha} = 0 \tag{2.69}$$

It means that σ can be considered as a function of $\varrho_{i\alpha}$, T. This is in agreement with experimental results (STEFAN, 1886; MUTAFTSCHIEV, 1982; SANCHEZ, 1983; KEENEY, HEICKLEN, 1979; RUSANOV, 1982) and represents, therefore, an additional verification of postulate 2.

Table 2.1:
Thermodynamic potentials of a heterogeneous system - cluster in the surrounding medium - for different thermodynamic constraints

Independent variables	Thermodynamic potential
T, V, n_i T = const. $T_\alpha = T_\beta = T$	$F = U - TS$ (Helmholtz free energy) $F_\alpha = -p_\alpha V_\alpha + \sum \mu_{i\alpha} n_{i\alpha}$ $F_\beta = -p_\beta V_\beta + \sum \mu_{i\beta} n_{i\beta}$; $F_0 = \sigma A + \sum \mu_{io} n_{io}$ $dF = -p_\alpha dV_\alpha + \sum \mu_{i\alpha} dn_{i\alpha} - p_\beta dV_\beta + \sum \mu_{i\beta} dn_{i\beta} +$ $+ \sigma dA + \sum \mu_{io} dn_{io}$
T, p, n_i T = const.	$G = U - TS + pV$; $p = p_\beta$ (Gibbs free energy) $G_\alpha = \sum \mu_{i\alpha} n_{i\alpha}$; $G_\beta = \sum \mu_{i\beta} n_{i\beta}$ $G_0 = V_\alpha (p_\beta - p_\alpha) + \sigma A + \sum \mu_{io} n_{io}$ $dG = (p_\beta - p_\alpha) dV_\alpha + \sum \mu_{i\alpha} dn_{i\alpha} + \sum \mu_{i\beta} dn_{i\beta} +$ $+ \sigma dA + \sum \mu_{io} dn_{io}$
S, p, n_i	$H = U + pV$; $p = p_\beta$ (enthalpy) $H_\alpha = T_\alpha S_\alpha + \sum \mu_{i\alpha} n_{i\alpha}$ $H_\beta = T_\beta S_\beta + \sum \mu_{i\beta} n_{i\beta}$ $H_0 = T_0 S_0 + \sum \mu_{io} n_{io} + \sigma A + V_\alpha (p_\beta - p_\alpha)$ $dH = T_\alpha dS_\alpha + V_\alpha dp_\alpha + \sum \mu_{i\alpha} dn_{i\alpha} + T_\beta dS_\beta +$ $+ V_\beta dp_\beta + \sum \mu_{i\beta} dn_{i\beta} + T_0 dS_0 + V_\alpha d(p_\beta - p_\alpha) +$ $+ (p_\beta - p_\alpha) dV_\alpha + \sigma dA + \sum \mu_{io} dn_{io}$
S, V, n_i	$U = U_\alpha + U_\beta + U_0$ (inner energy) $U_\alpha = T_\alpha S_\alpha - p_\alpha V_\alpha + \sum \mu_{i\alpha} n_{i\alpha}$ $U_\beta = T_\beta S_\beta - p_\beta V_\beta + \sum \mu_{i\beta} n_{i\beta}$ $U_0 = T_0 S_0 + \sigma A + \sum \mu_{io} n_{io}$ $dU = T_\alpha dS_\alpha - p_\alpha dV_\alpha + \sum \mu_{i\alpha} dn_{i\alpha} + T_\beta dS_\beta -$ $- p_\beta dV_\beta + \sum \mu_{i\beta} dn_{i\beta} + T_0 dS_0 + \sigma dA +$ $+ \sum \mu_{io} dn_{io}$

3. Thermodynamics and Nucleation in Finite Systems

3.1. The Work of Formation of Clusters

In metastable initially homogeneous systems due to fluctuation processes clusters of a new phase can be formed. If such a heterogeneous
state - cluster in the medium - is generated by some manipulations on
the system, then some work has to be done from outside. The minimum
value of the work needed to create the cluster is called the work of
formation of the cluster, which can be calculated by eq. (2.68).

Apart from the cases discussed above we restrict ourselves now to
isothermal processes and consider in the following two relevant constraints, either

$$T = \text{const}, \quad n_i = \text{const}, \quad V = \text{const} \tag{3.1}$$

$$\text{or} \quad T = \text{const}, \quad n_i = \text{const}, \quad p = \text{const} \tag{3.2}$$

For the constraints (3.1) the work of formation of clusters is given
by

$$W = F_{het} - F_{hom} = \Delta F \tag{3.3}$$

If the formation of s different clusters with a higher compared with
the surrounding medium density is considered, eq. (3.3) reads:

$$W = \sum_{l=1}^{s} \left\{ (p_\beta - p_\alpha^{(1)}) V_\alpha^{(1)} + \sum_{i=1}^{k} (\mu_{i\alpha}^{(1)} - \mu_{i\beta}) \tilde{n}_{i\alpha}^{(1)} + \sigma_\alpha^{(1)} A^{(1)} \right\} +$$
$$+ V(p - p_\beta) + \sum_{i=1}^{k} (\mu_{i\beta} - \mu_i) n_i \tag{3.4}$$

p and μ_i are the pressure and the chemical potential in the homogeneous initial state.

Considering the constraints (3.2) we obtain $W = \Delta G$. In this case the
term $V(p - p_\beta)$ is absent in the formulae for the work of formation of
clusters (3.4) because of the condition $p_\beta = p$.

For an illustration of the general results we investigate the formation of only one drop in a one-component supersaturated gas under isothermal conditions. If we assume that the surface tension does not de-
pend on the curvature ($n_0 = 0$), that means $\tilde{n}_\alpha$ is replaced by n_α, eq.
(3.4) is reduced to

$$\Delta F = (p_\beta - p_\alpha) V_\alpha + (\mu_\alpha - \mu_\beta) n_\alpha + \sigma A +$$
$$+ V(p - p_\beta) + (\mu_\beta - \mu) n \tag{3.5}$$

For the isobaric-isothermal case it holds:

$$\Delta G = (p - p_\alpha) \, V_\alpha + \sigma A + (\mu_\alpha - \mu) \, n_\alpha \qquad (3.6)$$

To calculate ΔF and ΔG the vapour is assumed to be a perfect gas. The initial supersaturation of the homogeneous state is then given by

$$y_0 = \frac{p}{p'(T)} = \frac{n \, RT}{p' \cdot V} \qquad (3.7)$$

As a reference pressure the equilibrium pressure p' for the coexistence of the liquid and the gas at a planar interface is chosen. Due to the constraints (3.2) the supersaturation of the system is always constant; while for the constraints (3.1) the actual supersaturation of the system changes because of the formation of the droplet. This fact will be discussed later.

We assume further incompressibility of the liquid, that means:

$$\rho_\alpha = n_\alpha / V_\alpha = \text{const.} \qquad (3.8)$$

The chemical potential of the incompressible drop respectively the perfect vapour is expressed as follows:

$$\mu_\alpha(p_\alpha, T) = \mu'(p', T) + \frac{1}{\rho_\alpha} (p_\alpha - p'(T)) \qquad (3.9)$$

$$\mu_\beta(p_\beta, T) = \mu'(p', T) + R_g T \ln p_\beta / p'(T) \qquad (3.10)$$

With these assumptions we obtain for the work of formation of one drop for the given constraints:

$$\Delta F = (p_\beta - p' - \rho_\alpha R_g T \ln p_\beta / p') \, V_\alpha + \sigma A \qquad (3.11)$$

$$+ (p - p_\beta) \, V + nR_0 T (\ln p_\beta / p' - \ln y_0)$$

$$\Delta G = (p - p' - \rho_\alpha R_g T \ln y_0) \, V_\alpha + \sigma A \qquad (3.12)$$

Note, that p_β for isochoric-isothermal constraints is also a function of the size of the drop, since

$$p_\beta = p_\beta(\rho_\beta, T); \qquad \rho_\beta = \frac{n_\beta}{V_\beta} = \frac{n - n_\alpha}{V - V_\alpha} = \frac{n - \rho_\alpha V_\alpha}{V - V_\alpha} \qquad (3.13)$$

Supposing a spherical drop with the radius r_α, the change of the free energy, ΔF, resp. the Gibbs free energy, ΔG, are presented in Figs. 3.1 and 3.2 for different values of the initial supersaturation y_0.

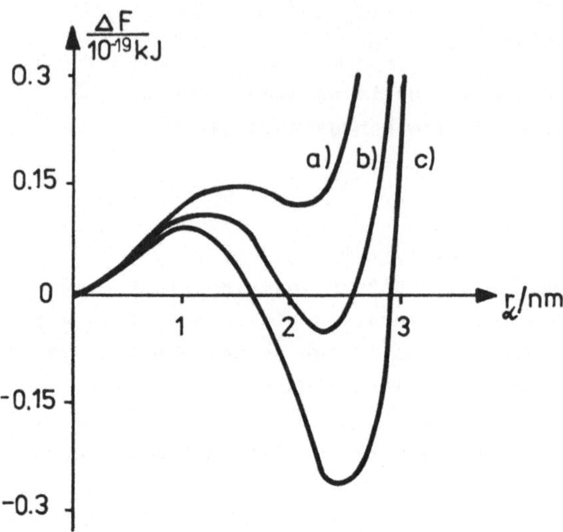

Fig. 3.1
Change of the free energy, ΔF, vs. radius of the drop for different values of the supersaturation y_o: a) y_o = 2.0; b) y_o = 2.2; c) y_o = 2.4
vapour: ethanol, n = $3 \cdot 10^{-19}$ mol, T = 312.35 K
(From: SCHWEITZER et al., 1984a)

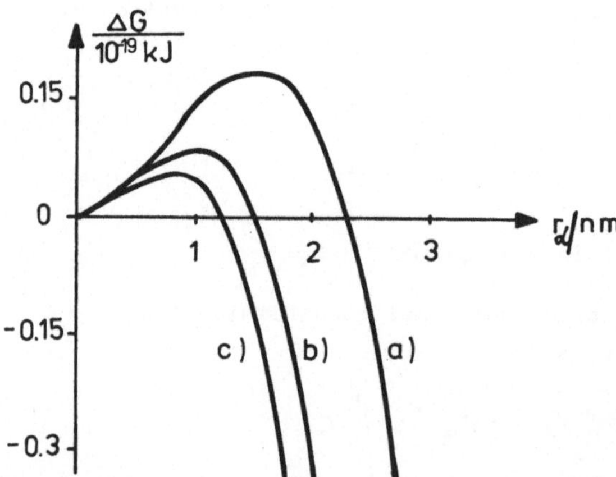

Fig. 3.2
Change of the Gibbs free energy, ΔG, vs. radius of the drop for different values of y_o: a) y_o = 1.8; b) y_o = 2.4; c) y_o = 3.0
vapour: ethanol, n = $3 \cdot 10^{-19}$ mol, T = 312.35 K
(From: SCHWEITZER et al., 1984a)

Fig. 3.1 shows that under isochoric-isothermal conditions the function ΔF has two extremum states. The maximum of ΔF corresponds to a critical droplet size, but the minimum of ΔF characterizes the stable state, the coexistence between the drop and the vapour. This stable state is caused by the change of the medium due to the formation of the cluster. The depletion of the medium is a typical effect for the nucleation in finite systems.

For isochoric constraints a critical respectively a stable state for the drop only exist in a certain range of the supersaturation. If y_0 is less than a critical value y_c, than the free energy only increases with an increasing droplet size and a phase transition cannot proceed via the mechanism of homogeneous nucleation. For a detailed discussion of this fact see Chapter 3.3.

For isobaric constraints only one extremum, a maximum, of ΔG is found as presented in Fig. 3.2. Omitting the condition of an incompressible cluster phase the maxima of ΔG and ΔF correspond to unstable equilibrium states of saddle-point type, as will be shown in Chapter 3.2. Both the radii of the critical and stable drop and the curves for the thermodynamic potentials depend on the initial supersaturation. Generally, higher initial supersaturations decrease the nucleation barrier and the critical droplet size.

Because of the condition of constant pressure the formation of the drop in isobaric-isothermal systems does not result in a depletion of the surrounding medium; this case is equivalent, therefore, to the case of formation of drops in an infinite system as discussed in Chapter 1.

The critical cluster radius under isobaric conditions is given by:

$$\frac{\partial \Delta G}{\partial r_\alpha} = 0 \; ; \quad r_c = \frac{2\sigma}{\varrho_\alpha R_g T \ln y_0 - (p-p')} \tag{3.14}$$

Substitution of r_c into eq. (3.12) results in

$$\Delta G_c = \frac{1}{3} \sigma A_c \qquad \text{with} \qquad A_c = 4\pi r_c^2 \tag{3.15}$$

respectively approximately in

$$\Delta G_c = \frac{16\pi}{3} \frac{\sigma^3}{(\varrho_\alpha R_g T \ln y_0)^2} \tag{3.16}$$

With eqs. (3.14), (3.16) another expression of the change of the Gibbs free energy is obtained (KAISCHEW, pers. communication)

$$\Delta G = \Delta G_c \left\{ -2\left(\frac{r_\alpha}{r_c}\right)^3 + 3\left(\frac{r_\alpha}{r_c}\right)^2 \right\} \tag{3.17}$$

If the case of formation of s identical drops is considered, then eq. (3.17) reads:

$$\Delta G = s\,\Delta G_c \left\{ -2\left(\frac{r_\alpha}{r_c}\right)^3 + 3\left(\frac{r_\alpha}{r_c}\right)^2 \right\} \tag{3.18}$$

where r_c does not depend on the number of clusters and is determined by eq. (3.14), again. The function $\Delta G = \Delta G(r_\alpha/r_c)$ is presented in Fig. 3.3 for different values of the number of drops.

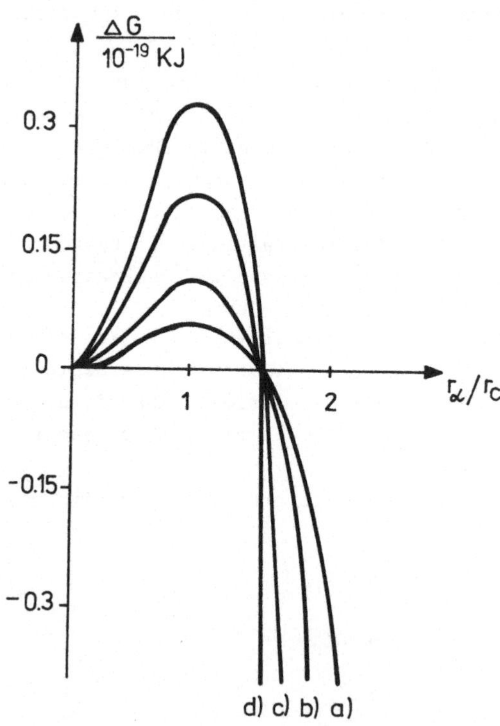

Fig. 3.3
Change of the Gibbs free energy ΔG as a function of the radius of the drops, for the case of formation of s identical drops. The position of the maximum and also the point of intersection with the r_α/r_c axis do not depend on the number of drops in the system
a) s = 60; b) s = 120; c) s = 360, d) s = 480; initial supersaturation: y_0 = 2.4
vapour: ethanol, $n = 3\cdot10^{-19}$ mol, T = 312.35 K

(From: SCHWEITZER et al., 1984a)

3.2. Equilibrium States and the Conditions for Stability of the Clusters

We now discuss the equilibrium conditions for the single drop in the one-component system. First we don't use the additional assumptions of a perfect gas and an incompressible liquid'.

For isochoric constraints the necessary equilibrium conditions then are obtained from eq. (3.4) by:

$$d(\Delta F) = 0 \qquad \text{with } l=1, \; i=1 \tag{3.19}$$

resulting in (SCHWEITZER, SCHMELZER, 1987):

$$f_1 = \mu_\alpha(\varrho_\alpha, T) - \mu_\beta(\varrho_\beta, T) = 0 \tag{3.20}$$

$$f_2 = -p_\alpha(\varrho_\alpha, T) + p_\beta(\varrho_\beta, T) + \frac{2\sigma(\varrho_\alpha, T)}{r_\alpha} = 0$$

The eqs. (3.20) determine two implicite functions

$$\tilde{n}_\alpha^{(i)} = \tilde{n}_\alpha^{(i)}(V_\alpha) \; ; \quad i=1,2$$

Their derivatives can be expressed by

$$\frac{\partial \tilde{n}_\alpha^{(i)}}{\partial V_\alpha} = - \frac{(\partial f_i/\partial V_\alpha)}{(\partial f_i/\partial n_\alpha)} \qquad i=1,2 \tag{3.21}$$

leading to (see also: SCHMELZER, SCHWEITZER, 1982)

$$\frac{\partial \tilde{n}_\alpha^{(1)}}{\partial V_\alpha} = \frac{\dfrac{1}{V_\alpha} \dfrac{\partial \mu_\alpha}{\partial \varrho_\alpha}\left[\dfrac{2\Gamma_0}{r_\alpha} + \varrho_\alpha\right] + \dfrac{\varrho_\beta}{V_\beta} \dfrac{\partial \mu_\beta}{\partial \varrho_\beta}}{\dfrac{1}{V_\alpha} \dfrac{\partial \mu_\alpha}{\partial \varrho_\alpha} + \dfrac{1}{V_\beta} \dfrac{\partial \mu_\beta}{\partial \varrho_\beta}} \tag{3.22}$$

$$\frac{\partial \tilde{n}_\alpha^{(2)}}{\partial V_\alpha} = \frac{\dfrac{1}{V_\alpha}\left(\varrho_\alpha + \dfrac{2\Gamma_0}{r_\alpha}\right)^2 \dfrac{\partial \mu_\alpha}{\partial \varrho_\alpha} + \dfrac{\varrho_\beta^2}{V_\beta} \dfrac{\partial \mu_\beta}{\partial \varrho_\beta} - \dfrac{\sigma}{2\pi r_\alpha^4}}{\dfrac{1}{V_\alpha}\left(\varrho_\alpha + \dfrac{2\Gamma_0}{r_\alpha}\right) \dfrac{\partial \mu_\alpha}{\partial \varrho_\alpha} + \dfrac{\varrho_\beta}{V_\beta} \dfrac{\partial \mu_\beta}{\partial \varrho_\beta}} \tag{3.23}$$

It follows, that the functions $\tilde{n}_\alpha^{(i)}$ show a behaviour as presented qualitatively in Fig. 3.4.

The points of intersection of the curves correspond to sets of parameters of the drop obeying the necessary equilibrium conditions. Moreover, it can be shown that the point of intersection for which the inequality

$$\frac{\partial \tilde{n}_\alpha^{(2)}}{\partial V_\alpha} > \frac{\partial \tilde{n}_\alpha^{(1)}}{\partial V_\alpha} \tag{3.24}$$

is fulfilled, corresponds to a stable equilibrium state, since equation (3.24) is equivalent to the necessary and sufficient stability condition (SCHMELZER, SCHWEITZER, 1982). On the other hand, the point of intersection, for which the inequality (3.25) holds

$$\frac{\partial \tilde{n}_\alpha^{(2)}}{\partial V_\alpha} < \frac{\partial \tilde{n}_\alpha^{(1)}}{\partial V_\alpha} \tag{3.25}$$

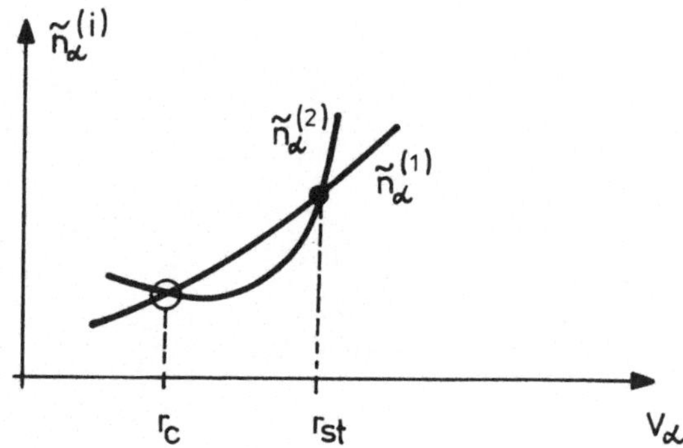

Fig. 3.4
Qualitative behaviour of the functions $\tilde{n}_\alpha^{(i)}$ determined by the equilibrium conditions (3.20) for isochoric-isothermal constraints in dependence on the radius of the drop
 o - unstable equilibrium state; ● - stable equilibrium state

(From: SCHMELZER, SCHWEITZER, 1982)

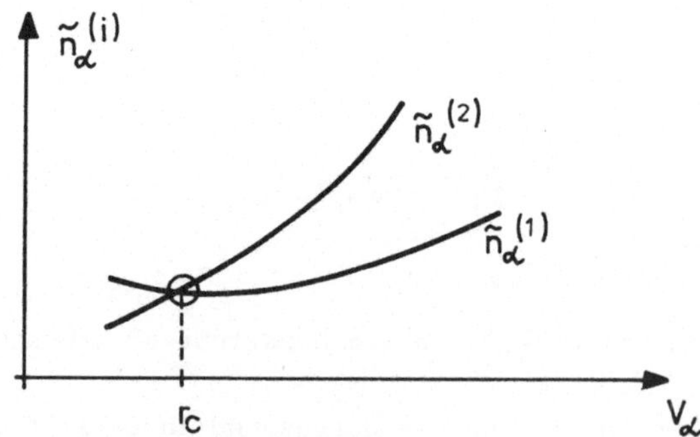

Fig. 3.5
Qualitative behaviour of the functions $\tilde{n}_\alpha^{(i)}$ determined by the equilibrium conditions (3.20) for isobaric-isothermal constraints in dependence on the radius of the drop
 o - unstable equilibrium state

(From: SCHWEITZER, 1983)

corresponds to the unstable equilibrium (critical state) (see also
KONOBEJEWSKI, 1934; HANSZEN, 1960; LACMANN, 1962).

Considering now isobaric-isothermal constraints the necessary equilibrium conditions (3.20) have to fulfilled with the restriction $p_\beta = p$
respectively $\varrho_\beta = \varrho$ = const. We can derive again two functions
$\tilde{n}_\alpha^{(1)}(V_\alpha)$; $\tilde{n}_\alpha^{(2)}(V_\alpha)$. For isobaric constraints their qualitative behaviour is presented in Fig. 3.5.

Fig. 3.5. shows that only one point of intersection exists, for which
the inequality (3.25) is valid. Therefore under isobaric conditions
the only equilibrium state is an unstable one.

Thus we conclude that the depletion of the medium in isochoric systems leads to a possible second equilibrium state for the drop, which
is a stable one. The two solutions of the equilibrium conditions may
coincide under special conditions discussed in Chapter 3.3.

Taking into account the assumptions of the incompressible drop (3.8)
and the perfect vapour, again, for isochoric constraints the necessary
equilibrium conditions result from eq. (3.5) as:

$$\frac{\partial \Delta F}{\partial r_\alpha} = - 4 \pi r_\alpha^2 \left\{ \varrho_\alpha R_g T \ln p_\beta/p' - (p_\beta-p') - \frac{2\sigma}{r_\alpha} \right\} = 0 \qquad (3.26)$$

Eq. (3.26) can be written in the form of a generalized Gibbs-Thomson
equation (SCHWEITZER, SCHMELZER, 1987):

$$\ln p_\beta/p'(T) - (p_\beta-p') = \frac{2\sigma}{r_\alpha \varrho_\alpha R_g T} \qquad (3.27)$$

If the term $(p_\beta-p')$ can be neglected compared with the first the
following relation holds for the critical respectively the stable
droplet radius

$$r_\alpha = \frac{2\sigma}{\varrho_\alpha R_g T \ln(p_\beta/p')} \qquad (3.28)$$

Let us note again, that p_β for the given constraints is a function
of the size of the single drop because of (3.13).

For isobaric conditions p_β is constant, that's why only one solution
of eq. (3.28) corresponding to the critical radius of the droplet is
obtained (compare also Fig. 3.2).

In the following we restrict ourselves to the discussion of the isochoric-isothermal case.

Taking into account the dependence of p_β on r_α , given by eq. (3.13),
eq. (3.28) can be transformed into eq. (3.29). A truncated Taylor expansion of the logarithm in eq. (3.28) leads to:

$$r^4 + R\ r + S = 0 \tag{3.29}$$

with the coefficients

$$R = -\ n\ \ln \frac{n\ R_g T}{p'\cdot V}\ \left(\frac{4\pi}{3}\ \rho_\alpha\right)^{-1} < 0 \tag{3.30}$$

$$S = n\ \left(\frac{2\sigma}{\rho_\alpha R_g T}\right)\ \left(\frac{4\pi}{3}\rho_\alpha\right)^{-1} > 0 \tag{3.31}$$

It has been shown that eq. (3.29) possesses only two positive solutions which depend on the thermodynamic parameters n, V and T (SCHWEITZER et al., 1984b). Fig. 3.6 shows these solutions in dependence on the system size.

The solutions of eq. (3.29) give the extremum droplet size. The smaller values correspond to the critical droplet size, the greater value to the stable droplet size in accordance with Fig. 3.1.

Both solutions of eq. (3.29) can coincide under certain thermodynamic conditions, as is demonstrated in Fig. 3.6. It was shown (SCHWEITZER et al., 1984a) that two solutions can be observed only if the inequation holds:

$$\left(\frac{R}{4}\right)^4 - \left(\frac{S}{3}\right)^3 > 0 \tag{3.32}$$

with R and S given by the eqs. (3.30), (3.31).

This inequality is the condition to find a stable droplet in the system. When both solutions coincide the stable droplet looses its stability. Therefore a nucleation process leading to a stable droplet

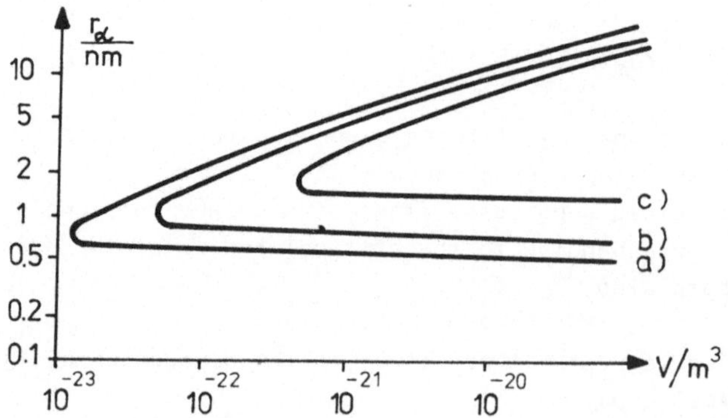

Fig. 3.6
Solutions of eq. (3.29) in dependence on the system volume V for different temperatures: a) T = 280 K, b) T = 290 K, c) T = 300 K vapor: ethanol, vapour density: n/V = 6.64 mol/m³

is possible in isochoric systems only if the thermodynamic parameters
N, V, T obey inequality (3.32). Some further conclusions are discussed
in the next Chapter.

3.3. Critical Thermodynamic Parameters for Nucleation in Finite Systems

Since the inequality (3.32) determines the boundaries for the possibility of a formation of a stable drop, eq. (3.33)

$$\left(\frac{R}{4}\right)^4 = \left(\frac{S}{3}\right)^3 \tag{3.33}$$

allows the calculation of the critical thermodynamic parameters for
homogeneous nucleation in finite systems.
Denoting by N the total number of particles in the system eqs. (3.30),
(3.31) and (3.33) yield

$$N\left(\ln\frac{Nk_BT}{p'V}\right)^4 = 4\left(\frac{4\pi}{3}\varrho_\alpha N_A\right)\left(\frac{4}{3}\frac{2\sigma}{\varrho_\alpha R_g T}\right)^3 \tag{3.34}$$

For fixed values of the temperature T and the volume V eq. (3.34) determines a critical number of particles N_c. Only for $N > N_c$ a stable
drop can be formed in the system. N_c is presented as a function of the
volume of the system for a constant temperature in Fig. 3.7.
On the other hand, if both V and N are kept constant, then eq. (3.34)
determines a critical temperature T_c. The notation critical temperature is meaningful, since only for $T < T_c$ a phase transition may occur.
Thus it rosemhles the common definition of a critical temperature used
for macroscopic (infinite) systems.
It was shown analytically, that the critical temperature is smaller
for a finite system compared with the macroscopic value (SCHWEITZER,
SCHIMANSKY-GEIER, 1987) with an increasing volume T_c tends to T_{cr} (macroscopic).
In Fig. 3.8 T_c is presented as a function of the system size V for
different (constant) initial supersaturations. It has to be taken into
account in the numerical calculations, that the saturation pressure p',
the surface tension σ and the density of the liquid ϱ_α are functions
of temperature. For the values see ZAHORANSKY (1982).
Thus it results, that nucleation processes in finite systems may proceed in a smaller interval of the thermodynamic parameters compared
with an infinite. This is illustrated also by the dependence of the
critical supersaturation y_c on the volume of the system (Fig. 3.9).
Again, only for $y_0 > y_c$ a phase transition may proceed via homogeneous

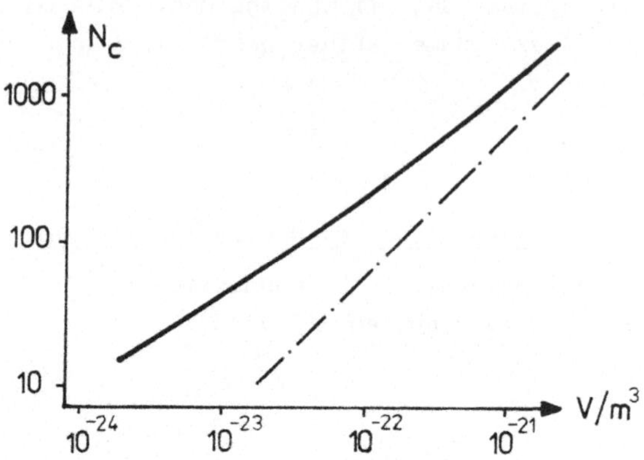

Fig. 3.7
Critical overall particle number N_c versus system volume V. Only for
$N > N_c$ a coexistence of the drop with the vapour is possible. The
dashed-dotted line gives the saturation particle number $N_o = p'V/k_B T$
vapour: ethanol, T = 280 K

(From: SCHWEITZER, SCHIMANSKY-GEIER, 1987)

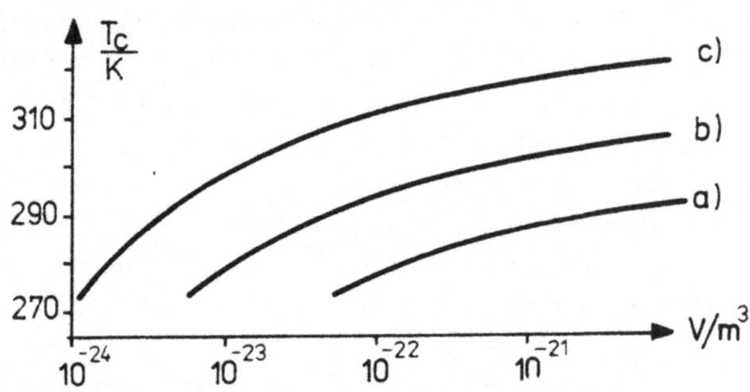

Fig. 3.8
Critical temperature T_c versus system volume V. If $T > T_c$ a stable
droplet is impossible to exist in the finite system. T_c is plotted
for various initial vapour densities: a) n/V = 3.32 mol/m³, b) n/V =
6.64 mol/m³, c) n/V = 13.28 mol/m³
vapour: ethanol

(From: SCHWEITZER, SCHIMANSKY-GEIER, 1987)

nucleation.

Using eqs. (3.7) and (3.34) we find for y_c (SCHWEITZER, 1986c)

$$y_c = \exp\left\{ \left(\frac{4\pi}{3} \varrho_\alpha N_A / \frac{N}{4}\right)^{1/4} \left(\frac{4}{3} \frac{2\sigma}{\varrho_\alpha R_g T}\right)^{3/4} \right\} \tag{3.35}$$

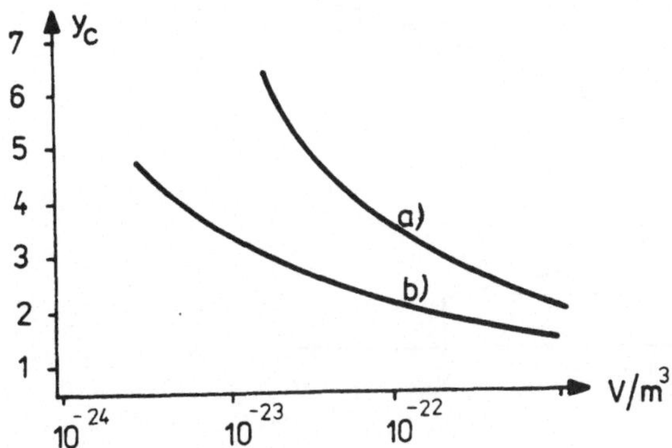

Fig. 3.9
Critical initial supersaturation y_c versus system volume V.
vapour: ethanol, a) T = 280 K, b) T = 312.35 K

(From: SCHWEITZER, SCHIMANSKY-GEIER, 1987)

If $y_0 = y_c$, the critical and the stable radius of the drop are just
coincide. The only solution of the Gibbs-Thomson equation (3.28) is
then given by:

$$r_{cs} = \frac{4}{3} \frac{2\sigma}{\rho_\alpha R_g T \ln y_c} \tag{3.36}$$

r_{cs} gives the smallest stable droplet size. The change of the free
energy ΔF (Fig. 3.1) in this case has only a point of inflexion and
no true extremum states:

$$\frac{\partial \Delta F}{\partial r_\alpha} = \frac{\partial^2 \Delta F}{\partial r_\alpha^2} = 0 \tag{3.37}$$

On the other hand, for initial supersaturations $y_0 > y_c$ the critical
radius decreases and the size of the stable drop increases with an
increasing supersaturation. Schematically this is shown in Fig. 3.10.
    For large initial supersaturations the critical radius becomes of
molecular size (r_0) the nucleation barrier disappears and nucleation
is replaced by spinodal decomposition. The value of the supersaturation, where nucleation goes over into spinodal decomposition, is indicated by y_{sd}, though, of course, no sharp boundary exists.
Consequently, phase transitions, which proceed by nucleation and growth,
may proceed for supersaturations $y_c < y_0 < y_{sd}$, only.

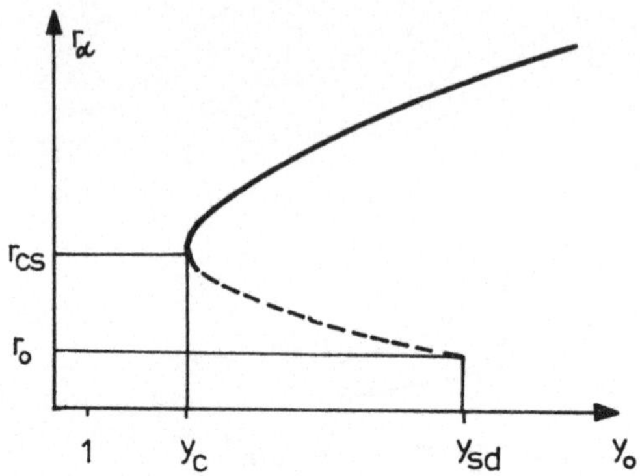

Fig. 3.10
Sketch of the stable droplet radius (solid line) and the critical droplet radius (dotted line) as a function of the initial supersaturation y_o
y_c is given by eq. (3.35), r_{cs} is given by eq. (3.37)
(From: SCHWEITZER, 1986c)

The results of the thermodynamic analysis, outlined here, are applicable to an explanation of the findings in computer calculations (THOMPSON et al., 1984) that stable drops in finite systems may not be formed though the vapour pressure is greater compared with the equilibrium pressure for a liquid-vapour system and a planar interface. This effect can be discussed either based on the size-dependence of the critical supersaturation or by the introduction of a critical volume V_c. For a given initial supersaturation y_o and temperature T we obtain from eq. (3.34) a value for V_c. For $V < V_c$ a stable drop cannot be formed.
Fig. 3.11 shows this critical system size in dependence on the initial supersaturation. In Table 3.1 some values of the critical system size are given together with the critical total particle number N_c, the radius of the drop r_{cs}, corresponding to the point of inflexion of the curve $\Delta F = \Delta F(r_\alpha)$, and the number of particles $N_{\alpha c}$ in such a drop.

Finally, we would like to mention, that the discussed effects may have some relevance for phase transitions in finite systems, e.g., in porous media.

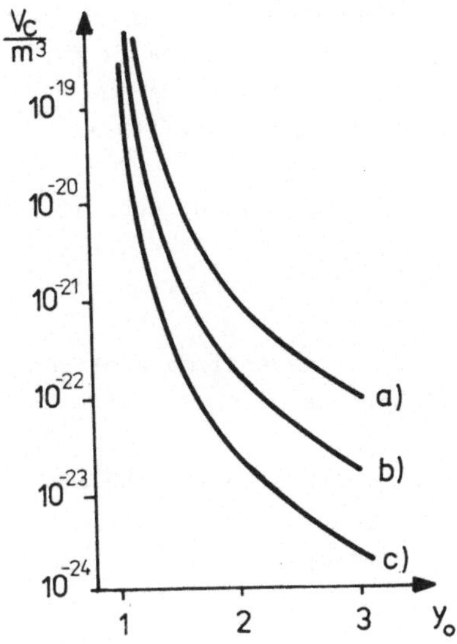

Fig. 3.11
Critical volume V_c of the system as a function of the initial supersaturation y_o for different values of the temperature, a) T =
288.83 K, b) T = 312.35 K, c) T = 343.71 K, vapour: ethanol

(From: SCHMELZER, SCHWEITZER, 1985)

Tab. 3.1
Critical volume V_c and related overall particle number N_c of the system for different temperatures and a constant initial supersaturation
y_o = 2. vapour: ethanol
For the given thermodynamic values only one solution of the equilibrium
condition (3.28) exists. r_{cs} (3.3) gives the radius of the drop in
this state and $N_{\alpha c}$ its particle number

(From: SCHWEITZER et al., 1984b)

T (K)	V_c (10^{-23} m^3)	N_c	$N_{\alpha c}$	r_{cs} (nm)
288.83	97.3	2262	391	2.059
312.35	16.4	1349	233	1.701
343.71	2.2	706	122	1.393

3.4. The Work of Formation of Critical Clusters

As it was already mentioned in Chapter 2.1. the work of formation of a critical cluster $W^{(e)}$ determines to a large extent the nucleation rate. Therefore, it is of interest to analyze, how finite size effects modify $W^{(e)}$.

As we will show two effects have to be taken into account. First, additional terms, compared with eq. (2.1), appear in the expression for $W^{(e)}$, second, depletion effects lead to a variation of the values of the parameters of the critical cluster.

The analysis is carried out here for the general case of one cluster in a k-component system. As an illustration, special cases are considered afterwards.

Considering isochoric-isothermal constraints (3.1), the work of formation of clusters is generally given by eq. (3.4).

Taking into account the necessary thermodynamic equilibrium conditions (2.66) eqs. (3.3), (3.4) yield

$$W^{(e)} = \frac{1}{3}\, \sigma A + V(p-p_\beta) + \sum_{i=1}^{k} (\mu_{i\beta} - \mu_i)\, n_i \tag{3.38}$$

In the same way we get for other constraints (see: SCHWEITZER, 1983; SCHMELZER , 1985b,c; SCHMELZER, SCHWEITZER, 1987b)

$$\Delta G^{(e)} = \frac{1}{3}\, \sigma A + \sum_{i=1}^{k}(\mu_{i\beta} - \mu_i)\, n_i$$

$$p = p_\beta = \text{const.} \qquad T = \text{const.} \qquad n_i = \text{const.}$$

$$\Delta U^{(e)} = \frac{1}{3}\, \sigma A + \sum_{i=1}^{k}(\mu_{i\beta} - \mu_i)\, n_i + (T_\beta - T)\, S + (p-p_\beta)\, V \tag{3.39}$$

$$V = \text{const.} \qquad S = \text{const.} \qquad n_i = \text{const.}$$

$$\Delta H^{(e)} = \frac{1}{3}\, \sigma A + \sum_{i=1}^{k}(\mu_{i\beta} - \mu_i)\, n_i + (T_\beta - T)\, S$$

$$p = p_\beta = \text{const.} \qquad T = \text{const.} \qquad n_i = \text{const.}$$

From eqs. (3.38) - (3.39) it can be obtained, that independent of the thermodynamic constraints the work of formation of the critical clusters can be expressed by the term W_{Gibbs}

$$W_{Gibbs} = \frac{1}{3}\, \sigma A \tag{3.40}$$

and a correction term ΔW. The correction term is due to the depletion of the surrounding the clusters medium as a result of the formation of the cluster. If depletion effects can be neglected, then ΔW is equal to zero. This is the case for large systems or, e.g., for one-component systems, if $p = p_\beta = $ const. and $T = $ const. are chosen as the thermodynamic constraints.

The problem of the calculation of the work of formation of critical clusters for different thermodynamic constraints was formulated already by RUSANOV (1967, 1978). We would like to show now that the correction term ΔW is less or equal to zero. This result is valid independent of the thermodynamic constraints (SCHWEITZER, 1983; SCHMELZER, SCHWEITZER, 1985; SCHMELZER, 1985b,c).

Here we consider again the thermodynamic constraints given by eq. (3.1). In this case ΔW can be expressed by

$$\Delta W = (p-p_\beta) \, V + \sum_{i=1}^{k} (\mu_{i\beta} - \mu_i) \tag{3.41}$$

The changes of the state of the medium due to the formation of a critical cluster are, in general, small and so we can expand p_β and $\mu_{i\beta}$ in a Taylor series. Denoting the independent variables for the time by x_i, $i=1,2,\ldots,k+1$ ($x_i=n_i$, $i=1,2,\ldots,k$; $x_{k+1}=V$) and their variations by Δx_i the first terms of the expansion of p_β and $\mu_{i\beta}$ are:

$$p_\beta = p(x_1,x_2,\ldots,x_{k+1}) + \sum_{l=1}^{k+1} \frac{\partial p}{\partial x_1} \Delta x_1 + \frac{1}{2}\sum_{l,m=1}^{k+1} \frac{\partial^2 p}{\partial x_1 \partial x_m} \, \Delta x_1 \, \Delta x_m \tag{3.42}$$

$$\mu_{i\beta} = \mu_i(x_1,\ldots,x_{k+1}) + \sum_{l=1}^{k+1} \frac{\partial \mu_i}{\partial x_1} \Delta x_1 + \frac{1}{2}\sum_{l,m=1}^{k+1} \frac{\partial^2 \mu}{\partial x_1 \partial x_m} \, \Delta x_1 \, \Delta x_m \tag{3.43}$$

For the homogeneous initial state the Gibbs-Duhem relation holds, which can be written as

$$- V \frac{\partial p}{\partial x_1} + \sum_{i=1}^{k} n_i \frac{\partial \mu_i}{\partial x_1} = 0 \tag{3.44}$$

Eqs. (3.41) - (3.44) result in

$$\Delta W = - \frac{1}{2}\left\{ \sum_{l,m=1}^{k+1} \left[\frac{\partial^2 p}{\partial x_1 \partial x_m} V - \sum_{i=1}^{k} n_i \frac{\partial^2 \mu_i}{\partial x_1 \partial x_m} \right] \Delta x_1 \Delta x_m \right\} \tag{3.45}$$

By a derivation of eq. (3.44) with respect to x_m one obtains

$$- \frac{\partial V}{\partial x_m} \frac{\partial p}{\partial x_1} - V \frac{\partial^2 p}{\partial x_1 \partial x_m} + \sum_{i=1}^{k} \frac{\partial n_i}{\partial x_m} \frac{\partial \mu_i}{\partial x_1} + \sum_{i=1}^{k} n_i \frac{\partial^2 \mu_i}{\partial x_1 \partial x_m} = 0 \tag{3.46}$$

and, finally, we get for ΔW

$$\Delta W = - \frac{1}{2} \sum_{l,m=1}^{k+1} \Delta x_1 \Delta x_m \left\{ - \frac{\partial V}{\partial x_m} \frac{\partial p}{\partial x_1} + \sum_{i=1}^{k} \frac{\partial n_i}{\partial x_m} \frac{\partial \mu_i}{\partial x_1} \right\} \qquad (3.47)$$

On the other hand, from the general thermodynamic equilibrium condition eq. (2.57) the following inequality can be derived:

$$\sum_{l,m=1}^{k+1} \delta x_1 \, \delta x_m \left\{ - \frac{\partial V}{\partial x_m} \frac{\partial p}{\partial x_1} + \sum_{i=1}^{k} \frac{\partial n_i}{\partial x_m} \frac{\partial \mu_i}{\partial x_1} \right\} > 0 \qquad (3.48)$$

A comparison of the eqs. (3.47) and (3.48) shows, that ΔW is a negative definite quadratic form of the variations of the state of the medium due to the formation of a critical cluster

$$\Delta W \leq 0 \qquad (3.49)$$

To illustrate the results given above we now consider the case of one incompressible drop in a one-component perfect vapour. With μ_β respectively μ given by eq. (3.10) we get the work of formation of the critical drop from eq. (3.38) in the form:

$$W^{(e)} = \frac{1}{3} \sigma A - V(p-p_\beta) - n R_g T (\ln p_\beta/p' - \ln y_0) \qquad (3.50)$$

Note again, that the surface area $A = 4 \pi r_\alpha^2$ and p_β (3.13) both depend on the radius r_α of the drop. For the critical drop therefore r_α has to satisfy the Gibbs-Thomson equation (3.27). The eqs. (3.27) and (3.50) with eq. (3.13) form a system of equations which has to be solved numerically.

In the Figs. 3.11, 3.12 and 3.13 there are presented the work of formation of critical clusters $W^{(e)}$ (3.50), W_{Gibbs} (3.40) and the correction term ΔW (3.41) in dependence on the total volume of the system. Since the Gibbs-Thomson equation for isochoric-isothermal systems has two solutions, the critical and the stable droplet size, one part of the curve refers to the formation of the critical drop, the other to the formation of the stable drop.

The Figs. 3.12, 3.13 and 3.14 show that $W^{(e)}$, W_{Gibbs} and ΔW behave qualitatively different as a function of the volume of the system in dependence whether the stable or the critical drop is considered. For the unstable (critical) drop the free energy of formation decreases with an increasing system size if the initial supersaturation is kept constant. This can be understood only if in addition to ΔW the dependence of the parameters of the critical cluster on depletion effects is taken into account (see next Chapter).

The correction term ΔW is small for the critical clusters, it tends to zero with an increasing system size.

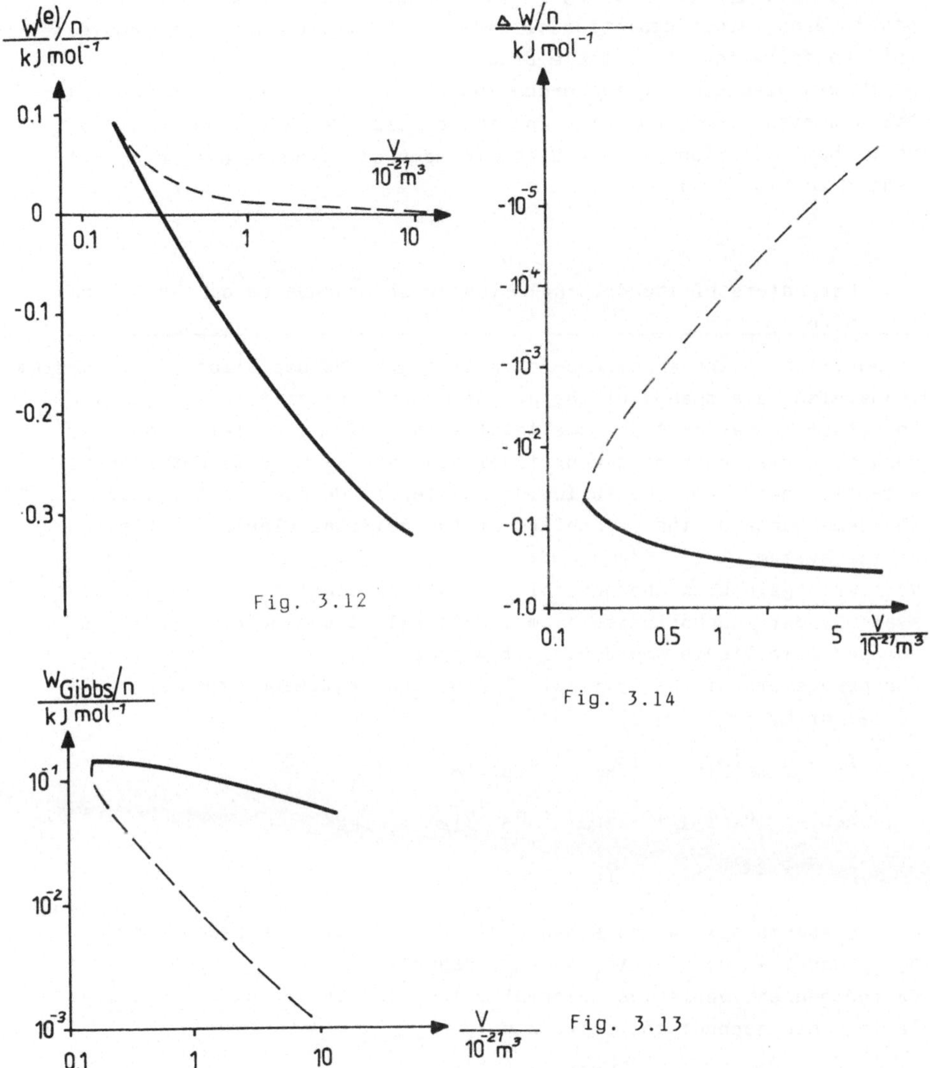

Figs. 3.12 - 3.14:
The molar work of formation of a critical respectively a stable drop $W^{(e)}/n$ (eq. 3.50) (Fig. 3.12), the molar surface energy W_{Gibbs}/n (eq. 3.40) (Fig. 3.13) and the molar correction term $\Delta W/n$ (eq. 3.41) (Fig. 3.14) in dependence on the total volume of the system. The dashed lines give the curves for the formation of a critical drop, the solid lines give the curves for the formation of the stable drop for a constant supersaturation. vapour: ethanol, y_o = 2, T = 312.35 K
(From: SCHMELZER and SCHWEITZER, 1985)

In contrast, ΔW gives a significant contribution to $W^{(e)}$ for the stable drop, since depletion effects of the medium are a prerequisite for the formation of a stable drop.

As it was discussed in the preceding Chapter the curves for the critical and stable drop coincide for the critical size of the system V_c, where both solutions of the Gibbs-Thomson equation become identical (see also Fig. 3.6).

3.5. Parameters of the Critical Cluster in Dependence on the System Size

In addition to the appearence of a term ΔW the depletion of the medium leads also to a change of the parameters of the critical clusters. Therefore, to account for the total effect of the depletion on the work of formation W of the critical clusters we have to consider both effects. That's why the following chapter is devoted to the study of the dependence of the parameters of the critical cluster on the size of the system.

We start again with the general case of one cluster in a k-component system under isochoric-isothermal constraints and do not assume an incompressible liquid and a perfect vapour.

The parameters of the critical cluster are determined then by eq. (2.66) or by

$$f_i = \mu_{i\alpha}(\rho_{1\alpha},\dots,\rho_{k\alpha}) - \mu_{i\beta}(\rho_{1\beta},\dots,\rho_{k\beta}) = 0$$

$$f_{k+1} = -p_\alpha(\rho_{1\alpha},\dots,\rho_{k\alpha}) + p_\beta(\rho_{1\beta},\dots,\rho_{k\beta}) +$$

$$+ 2\sigma(\rho_{1\alpha},\dots,\rho_{k\alpha})/r_\alpha = 0 \tag{3.51}$$

Let us assume now, we have two different systems, characterized by V, n_i, T and $V + \Delta V$, $n_i + \Delta n_i$ and T, respectively.

As independent variables describing the cluster V_α and $\tilde{n}_{i\alpha}$ are used. Taking into account the relations

$$\rho_{i\alpha} = \frac{n_{i\alpha}}{V_\alpha} \; ; \; \rho_{i\beta} = \frac{n_{i\beta}}{V_\beta} \; ; \; n_i = n_{i\alpha}+n_{i\beta}+n_{io}; \; V = V_\beta +V_\alpha \tag{3.52}$$

and in addition

$$\frac{\partial\sigma}{\partial\rho_{j\alpha}} = -\sum_{i=1}^{k}\Gamma_{io}\frac{\partial\mu_{i\alpha}}{\partial\rho_{j\alpha}} \qquad \Gamma_{io} = \frac{n_{io}}{A} \tag{3.53}$$

resulting from the Gibbs adsorption isotherm, we get for fixed values of n_i, V, T

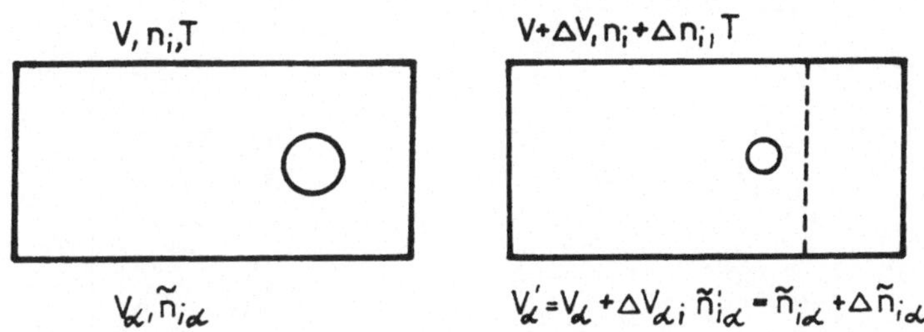

Fig. 3.15
Model used for the calculation of the dependence of the critical parameters V_α, $\tilde{n}_{i\alpha}$ of the cluster on the depletion of the medium due to the formation of the cluster

$$\tilde{n}_{i\alpha} = \tilde{n}_{i\alpha}(V, n_1, \ldots, n_k) \qquad V_\alpha = V_\alpha(V, n_1, \ldots, n_k) \qquad (3.54)$$

If the size of the system is changed the variables ϱ_i, T being unvaried in the initial state, the parameters of the critical cluster are given by

$$\tilde{n}'_{i\alpha} = \tilde{n}_{i\alpha}(V + \Delta V,\ n_1 + \Delta n_1, \ldots, n_k + \Delta n_k)$$
$$V'_\alpha = V_\alpha(V + \Delta V, n_1 + \Delta n_1, \ldots, n_k + \Delta n_k) \qquad (3.55)$$

Starting from these equations it is possible to calculate the variations $\Delta \tilde{n}_{i\alpha} = \tilde{n}'_{i\alpha} - \tilde{n}_{i\alpha}$, $\Delta V_\alpha = V'_\alpha - V_\alpha$ of the parameters of the critical clusters due to the change of the size of the system and, in general, also of the temperature (SCHMELZER, SCHWEITZER, 1985; SCHMELZER, 1985a; SCHWEITZER, 1987). As the result we obtain

$$\Delta V_\alpha = - \frac{\Delta V}{JV_\beta} \begin{vmatrix} \dfrac{\partial f_1}{\partial \tilde{n}_{1\alpha}} & \dfrac{\partial f_1}{\partial \tilde{n}_{2\alpha}} & \cdots & \dfrac{\partial f_1}{\partial \tilde{n}_{k\alpha}} & \sum\limits_{e=1}^{k} (\varrho_{e\beta} - \varrho_e) \dfrac{\partial \mu_{1\beta}}{\partial \varrho_{e\beta}} \\[2ex] \cdots & \cdots & \cdots & \cdots & \cdots \\[2ex] \dfrac{\partial f_k}{\partial \tilde{n}_{1\alpha}} & \dfrac{\partial f_k}{\tilde{n}_{2\alpha}} & \cdots & \dfrac{\partial f_k}{\partial \tilde{n}_{k\alpha}} & \sum\limits_{e=1}^{k} (\varrho_{e\beta} - \varrho_e) \dfrac{\partial \mu_{k\beta}}{\partial \varrho_{e\beta}} \\[2ex] \dfrac{\partial f_{k+1}}{\partial \tilde{n}_{1\alpha}} & \dfrac{\partial f_{k+1}}{\partial \tilde{n}_{2\alpha}} & \cdots & \dfrac{\partial f_{k+1}}{\partial \tilde{n}_k} & -\sum\limits_{e=1}^{k} (\varrho_{e\beta} - \varrho_e) \dfrac{\partial p_\beta}{\partial \varrho_{e\beta}} \end{vmatrix} \qquad (3.56)$$

$$\Delta \tilde{n}_{1\alpha} = - \frac{\Delta V}{J V_\beta} \begin{vmatrix} \sum\limits_{e=1}^{k} (\varrho_{e\beta} - \varrho_e) \dfrac{\partial \mu_{1\beta}}{\partial \varrho_{e\beta}} \dfrac{\partial f_1}{\partial \tilde{n}_{2\alpha}} & \cdots & \dfrac{\partial f_1}{\partial \tilde{n}_{k\alpha}} & \dfrac{\partial f_1}{\partial V_\alpha} \\ \cdots \cdots \cdots \cdots \cdots \cdots \cdots \\ \sum\limits_{e=1}^{k} (\varrho_{e\beta} - \varrho_e) \dfrac{\partial \mu_k}{\partial \varrho_{e\beta}} \dfrac{\partial f_k}{\partial \tilde{n}_{2\alpha}} & \cdots & \dfrac{\partial f_k}{\partial \tilde{n}_{k\alpha}} & \dfrac{\partial f_k}{\partial V_\alpha} \\ - \sum\limits_{e=1}^{k} (\varrho_{e\beta} - \varrho_e) \dfrac{\partial p_\beta}{\partial \varrho_{e\beta}} \dfrac{\partial f_{k+1}}{\partial \tilde{n}_{2\alpha}} & \cdots & \dfrac{\partial f_{k+1}}{\partial \tilde{n}_{k\alpha}} & \dfrac{\partial f_{k+1}}{\partial V_\alpha} \end{vmatrix} \tag{3.57}$$

Here the Jacobideterminant J is defined by (FICHTENHOLZ, 1969)

$$J = \frac{\partial (f_1, f_2, \ldots, f_{k+1})}{\partial (\tilde{n}_{1\alpha}, \tilde{n}_{2\alpha}, \ldots, V_\alpha)} = \begin{vmatrix} \dfrac{\partial f_1}{\partial \tilde{n}_{1\alpha}} \dfrac{\partial f_1}{\partial \tilde{n}_{2\alpha}} & \cdots & \dfrac{\partial f_1}{\partial \tilde{n}_{k\alpha}} \dfrac{\partial f_1}{\partial V_\alpha} \\ \dfrac{\partial f_2}{\partial \tilde{n}_{1\alpha}} \dfrac{\partial f_2}{\partial \tilde{n}_{2\alpha}} & \cdots & \dfrac{\partial f_2}{\partial \tilde{n}_{k\alpha}} \dfrac{\partial f_2}{\partial V_\alpha} \\ \cdots \cdots \cdots \cdots \cdots \cdots \cdots \\ \dfrac{\partial f_{k+1}}{\partial \tilde{n}_{1\alpha}} \dfrac{\partial f_{k+1}}{\partial \tilde{n}_{2\alpha}} & \cdots & \dfrac{\partial f_{k+1}}{\partial \tilde{n}_{k\alpha}} \dfrac{\partial f_{k+1}}{\partial V_\alpha} \end{vmatrix} \tag{3.58}$$

The expressions for $\Delta \tilde{n}_{i\alpha}$ can be obtained, if instead of the first column of the Jacobideterminant (3.58) the i-th column is replaced by the column consisting of the sums. Moreover, $\Delta \tilde{n}_{i\alpha}$ is with an accuracy to higher order terms equal to $\Delta n_{i\alpha}$.
The partial derivatives in eqs. (3.56) – (3.58) are given by (see SCHMELZER, 1985a; SCHMELZER, SCHWEITZER, 1987c)

$$\frac{\partial f_i}{\partial \tilde{n}_{1\alpha}} = \frac{1}{V_\alpha} \frac{\partial \mu_{i\alpha}}{\partial \varrho_{1\alpha}} + \frac{1}{V_\beta} \frac{\partial \mu_{i\beta}}{\partial \varrho_{1\beta}} \qquad i,1 = 1,2,\ldots,k$$

$$\frac{\partial f_i}{\partial V_\alpha} = - \sum_{l=1}^{k} \frac{1}{V_\alpha} \quad \frac{2 \Gamma_{1o}}{r_\alpha} + \varrho_{l\alpha} \frac{\partial \mu_{i\alpha}}{\partial \varrho_{1\alpha}} - \sum_{l=1}^{k} \frac{\varrho_{1\beta}}{V_\beta} \frac{\partial \mu_{i\beta}}{\partial \varrho_{1\beta}} \tag{3.59}$$

$$\frac{\partial f_{k+1}}{\partial \tilde{n}_{1\alpha}} = - \frac{1}{V_\alpha} \frac{\partial p_\alpha}{\partial \varrho_{1\alpha}} - \frac{1}{V_\beta} \frac{\partial p_\beta}{\partial \varrho_{1\beta}} + \frac{2}{r_\alpha} \frac{\partial \sigma}{\partial \varrho_{1\alpha}}$$

$$\frac{\partial f_{k+1}}{\partial V_\alpha} = \sum_{l=1}^{k} \frac{1}{V_\alpha} \frac{\partial p_\alpha}{\partial \varrho_{1\alpha}} \left[\frac{2 \Gamma_{1o}}{r_\alpha} + \varrho_{l\alpha} \right] + \sum_{l=1}^{k} \frac{\varrho_{l\beta}}{V_\beta} \frac{\partial p_\beta}{\partial \varrho_{1\beta}} - \frac{\sigma}{2\pi r_\alpha^4} \tag{3.60}$$

$$- \frac{2}{r_\alpha} \sum_{l=1}^{k} \frac{1}{V_\alpha} \frac{\partial \sigma}{\partial \varrho_{1\alpha}} \left[\frac{2 \Gamma_{1o}}{r_\alpha} + \varrho_{l\alpha} \right]$$

Further, the stability investigations show, that the Jacobideterminant J is greater than zero, if the heterogeneous state - cluster in the surrounding medium - is a stable equilibrium state. $J < 0$ corresponds to an unstable equilibrium state (critical cluster). So, in general, the variations of the parameters of the cluster will be different for stable and unstable equilibrium states, respectively.

Moreover, from eqs. (3.56) and (3.57) it follows, that $\Delta \tilde{n}_{i\alpha}$ and ΔV_α are linear functions of the variations of the state of the medium due to the formation of the critical and stable clusters. Therefore, the variations of the parameters of the critical clusters determine the change of the work of formation of the critical cluster due to depletion effects and not the correction term ΔW, which is a quadratic function of the relatively small variations of the state of the medium. Consequently, Gibbs' expression $W^{(e)} = \sigma A/3$ is a good approximation for the work of formation of the critical clusters independent of the thermodynamic constraints, if in the determination of the parameters of the cluster depletion effects are taken into account.

In particular, for one-component systems, we get from the general equations (3.56) - (3.57)

$$\Delta n_\alpha = \frac{\Delta V}{J} \frac{1}{V_\beta} \frac{\partial \mu_\beta}{\partial \rho_\beta} (\rho - \rho_\beta) \left[\frac{\rho_\alpha}{V_\alpha} \frac{\partial \mu_\alpha}{\partial \rho_\alpha} \left(\frac{2\Gamma_0}{r_\alpha} + \rho_\alpha - \rho_\beta \right) - \frac{\sigma}{2\pi r_\alpha^4} \right]$$

$$\Delta V_\alpha = \frac{\Delta V}{J} \frac{1}{V_\beta} \frac{\partial \mu_\beta}{\partial \rho_\beta} (\rho - \rho_\beta) \frac{1}{V_\alpha} \frac{\partial \mu_\alpha}{\partial \rho_\alpha} \left(\frac{2\Gamma_0}{r_\alpha} + \rho_\alpha - \rho_\beta \right) \qquad (3.61)$$

The change of the density of the i-th component can be expressed by

$$\Delta \rho_{i\alpha} = \frac{n_{i\alpha} + \Delta n_{i\alpha}}{V_\alpha + \Delta V_\alpha} - \frac{n_{i\alpha}}{V_\alpha} \simeq \frac{1}{V_\alpha} \left\{ \Delta n_{i\alpha} - \rho_{i\alpha} \Delta V_\alpha \right\}$$

leading for one-component systems to

$$\Delta \rho_\alpha = - \frac{\Delta V}{J} \frac{1}{V_\beta} \frac{\partial \mu_\beta}{\partial \rho_\beta} \frac{\rho - \rho_\beta}{V_\alpha} \frac{\sigma}{2\pi r_\alpha^4} \qquad (3.62)$$

So, with an increase of the volume of the system ($\Delta V > 0$) the volume of the critical cluster (e.g. drop) decreases ($J < 0$) and its density increases.
The volume of a cluster in stable equilibrium with the medium increases ($J > 0$) and its density decreases. Therefore, the density of a cluster increases with decreasing size of the cluster.

If we consider in contrast the formation of bubbles, clusters with a
lower compared with the surrounding medium density, then the following inequalities are fulfilled

$$\frac{2\Gamma_0}{r_\alpha} + \varrho_\alpha - \varrho_\beta < 0 \qquad \varrho < \varrho_\beta \qquad (3.63)$$

Again, the volume of a critical bubble decreases with an increasing
volume of the system, while the volume of the stable bubble increases
also. In contrast to the preceding case the density of the bubble decrease with a decreasing size of the bubble (see also SCHMELZER,
1986).

We finally summarize that the variation of the parameters of the critical cluster mainly determine the change of the work of formation
of such clusters due to the term W_{Gibbs} (eq. (3.40)) and not the term
ΔW. This was shown by explicit expressions for the dependence of
theses parameters on the depletion of the medium. The depletion leads
to an increase of the size and therefore of the work of formation of
the critical clusters.
For one-component systems under a constant external pressure (ϱ =
ϱ_β) the size of the critical clusters does not depend on the size of
the system, since no depletion effects occur.

3.6. Formation of a Droplet Ensemble in Finite Systems

If more than one cluster is formed in the system under isochoric-
isothermal constraints then we have to start with the general equation
(3.4) describing the work of formation of s different clusters in a
k-component system.
Let us restrict ourselves again to the one-component system and assume
further incompressible liquid clusters. Generally, the work of formation of the s drops is given then by (SCHMELZER, 1985b,c):

$$\Delta F = \sum_{l=1}^{s} \left\{ V_\alpha^{(1)} \left[(p_\beta - p') - \varrho_\alpha(\mu_\beta - \mu') \right] + \sigma^{(1)} A^{(1)} + \right.$$
$$+ (p - p_\beta) V + (\mu_\beta - \mu) n \qquad (3.64)$$

$\mu' = \mu'(p',T)$ denotes the chemical potential for the saturation
pressure p' at the temperature T.
The necessary equilibrium conditions can be fulfilled only, if all
drops have the same parameters. In the case of s identical drops eq.
(3.64) is reduced to

$$\Delta F = (p-p_\beta) \, V + (\mu_\beta-\mu) \, n +$$

$$+ s \, V_\alpha \left\{ (p_\beta-p') - \varrho_\alpha(\mu_\beta-\mu') \right\} + s \, \sigma A \tag{3.65}$$

The change of ΔF as a function of the common radius r_α of the drops
is presented in Fig. 3.16 for different numbers of clusters, the
vapour is considered as a perfect gas.

In contrast to the behaviour of the system under isobaric-isothermal
constraints (compare Fig. 3.3) here the positions of the extrema of
ΔF change with an increasing number of the drops. We obtain a critical number of drops, s_c, where the two extrema of the free energy
coincide. The critical number of drops and the related common radius
r_{cs} can be calculated by:

$$\left(\frac{\partial \Delta F}{\partial r_\alpha} \right)_s = \left(\frac{\partial^2 \Delta F}{\partial r_\alpha^2} \right)_s = 0 \tag{3.66}$$

The first derivation of ΔF determining the positions of the extrema
is given by:

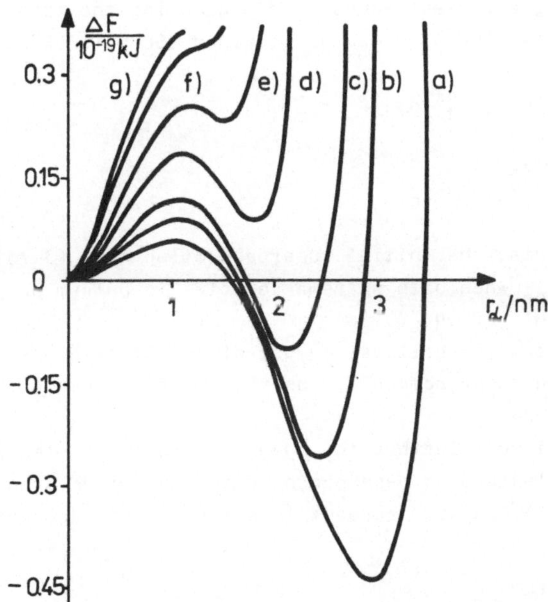

Fig. 3.16
Change of the free energy ΔF (3.65) versus the common droplet radius
r_α for different numbers of identical drops: a) s = 60, b) s = 90,
c) s = 120, d) s = 180, e) s = 240, f) s = 300, g) s = 360; initial
supersaturation: y_o = 2.4; vapour: ethanol, T= 312.35 K, $n=3\cdot10^{-19}$ mol

(From: SCHWEITZER et al., 1984a)

$$\left(\frac{\partial \Delta F}{\partial r_\alpha}\right)_s = -\,4\pi r_\alpha^2 s \left\{ \varrho_\alpha(\mu_\beta-\mu') - (p_\beta-p') - \frac{2\sigma}{r_\alpha} \right\} = 0 \tag{3.67}$$

leading to the generalized Gibbs-Thomson equation again:

$$\varrho_\alpha(\mu_\beta-\mu') - (p_\beta-p') = \frac{2\sigma}{r_\alpha} \tag{3.68}$$

In the case of a perfect vapour eq. (3.68) is reduced to eq. (3.27).
But now μ_β and p_β depend on the number of droplets too, because they
are functions of ϱ_β which is given, now, by:

$$\varrho_\beta = \varrho\,\frac{1 - s\,\dfrac{n_\alpha}{n}}{1 - s\,\dfrac{V_\alpha}{V}} \quad , \quad \varrho = \frac{n}{V} \tag{3.69}$$

That's why the values of the critical radius r_c and the radius of the
stable drop, r_s, both depend now on the number of drops present in
the system.

From the condition that both solutions of the eq. (3.68) coincide we
receive again the critical number of drops s_c and their common radius
r_{cs}. Assuming a perfect vapour and neglecting the term $(p_\beta - p')$ compared with the others it can be obtained (SCHWEITZER et al. 1984b)

$$s_c = \frac{n}{4}\;\ln y_0\;\left(\frac{4\pi}{3}\varrho_\alpha\right)^{-1}\;r_{sc}^{-3} \tag{3.70}$$

$$r_{cs} = \frac{4}{3}\;\frac{2\sigma}{\varrho_\alpha R_g T}\;\frac{1}{\ln y_0} \tag{3.71}$$

where y_0 denotes the initial supersaturation (eq. (3.7)).
The radius r_{cs} when both extrema coincide is determined in agreement
with the result of eq. (3.36).
Fig. 3.17 shows the critical number of identical drops for the finite system and their common radius r_{cs} in dependence on the initial
supersaturation.
We want to discuss further the change of radius of the critical and
the stable clusters in dependence on the number of drops. A derivation of eq. (3.68) with respect to s taking into account eq. (3.69)
leads to:

$$\frac{dr_\alpha}{ds} = -\,\frac{r_\alpha}{3s}\,\frac{1}{1+Z^{-1}} \qquad Z = -\,3\,\frac{\partial \mu_\beta}{\partial \varrho_\beta}\,\frac{r_\alpha(\varrho_\alpha-\varrho)^2\,s\,V_\alpha}{2\sigma\,(1-s\,\frac{V_\alpha}{V})^3\,V} \tag{3.72}$$

Further, since the second derivative of ΔF with respect to s reads

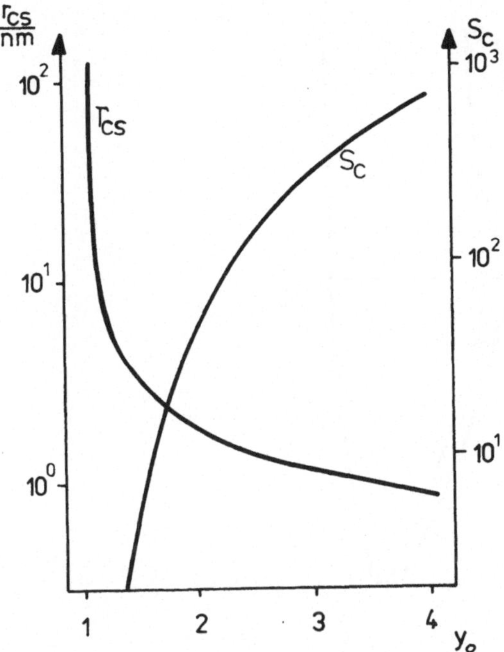

Fig. 3.17
Critical number of identical drops s_c (eq. (3.70)) and common radius r_{cs} (eq. (3.71) in dependence on the initial supersaturation y_o vapour: ethanol, T = 312.35 K, $n = 3 \cdot 10^{-19}$ mol

(From: SCHWEITZER et al., 1984b)

$$\left(\frac{\partial^2 \Delta F}{\partial r_\alpha^2} \right)_s = 8 \pi \sigma s (1 + Z) \tag{3.73}$$

we obtain from the extremum conditions

$$\left(\frac{\partial^2 \Delta F}{\partial r_\alpha^2} \right)_s < 0 \quad 1 + Z > 0 \quad \frac{dr_\alpha}{ds} > 0 \quad \text{for } r_\alpha = r_c \tag{3.74}$$

$$\left(\frac{\partial^2 \Delta F}{\partial r_\alpha^2} \right)_s > 0 \quad 1 + Z < 0 \quad \frac{dr_\alpha}{ds} < 0 \quad \text{for } r_\alpha = r_s$$

With an increasing number of drops r_c increases and r_s decreases (compare Fig. 3.15). For the point of inflexion of ΔF, given by the condition (3.66), it yields

$$Z = -1 \quad \frac{dr_\alpha}{ds} = 0 \quad \text{for } r_c = r_s = r_{cs} \tag{3.75}$$

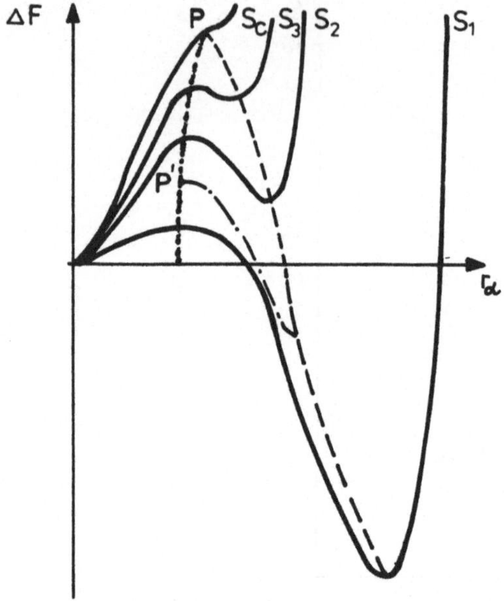

Fig. 3. 18
Change of the free energy ΔF versus droplet radius r_α for different numbers of identical drops $(s_1 < s_2 < s_3 < s_c)$ (schematic plot) (compare Fig. 3.16)

The variation of the extrema of ΔF with respect to s can be calculated from eq. (3.65), a derivation of this equation with respect to s results in

$$\frac{d\Delta F}{ds} = \left(\frac{\partial \Delta F}{\partial r_\alpha}\right)_s \frac{dr_\alpha}{ds} + \left(\frac{\partial \Delta F}{\partial s}\right)_{r_\alpha} = \frac{1}{3} \, \sigma \, A \tag{3.76}$$

The change of the extrema of ΔF with an increasing number of drops is presented in Fig. 3.16. Moreover, a derivation of the quantity $\Delta F/s$ with respect to s leads to

$$\frac{d}{ds}\left(\frac{\Delta F}{s}\right) = -\frac{1}{s^2}\left(\Delta F - s \frac{1}{3}\sigma A\right) = -\frac{\Delta W}{s^2} \geqslant 0 \tag{3.77}$$

Considering the case of a fixed number of drops and an increasing supersaturation the radius of the critical drops decreases and the radius of the stable drops increases (compare Fig. 3.1 for the case s = 1). This conclusion can be derived directly from eq. (3.78), which results from a derivation of eq. (3.67) with respect to the initial density ρ and eq. (3.69):

$$\left(\frac{dr_\alpha}{d\rho}\right)_s = \frac{\dfrac{\rho_\alpha - \rho_\beta}{\rho}}{\dfrac{1}{4\pi r_\alpha^2 s}} \quad \frac{\dfrac{\partial \mu_\beta}{\partial \rho_\beta} \dfrac{n - s n_\alpha}{V - s V_\alpha}}{\left(\dfrac{\partial^2 \Delta F}{\partial r_\alpha^2}\right)} \tag{3.78}$$

The results of the thermodynamic investigations, summarized in Fig. 3.18, lead to the distinction of, at least, three main different stages of first-order phase transitions in finite systems, a stage of nucleation (dotted curve), a stage of practically independent growth of the clusters, their number being nearly constant (dashed-dotted curve) and a stage of competitive growth of Ostwald ripening, leading to a decrease of the number of drops and an increase of their average size.
Moreover, since for $s > s_c$ ΔF is a monotonously increasing function of the radius, s_c can be considered as the maximum number of drops which can be formed by nucleation. In general, this maximum number will not be formed. Nucleation will be replaced earlier by the stage of independent growth, which starts at a point, indicated in Fig 3.18 by P'.
The thermodynamic results are used now as a basis for a kinetic description of first-order phase transitions, outlined in Chapters 4 - 6.

4. Kinetics of Phase Transitions in Finite Systems - A Stochastic Approach

4.1. Free Energy of the Cluster Distribution

After investigating the thermodynamics of the heterogeneous system
which has been established due to the first order phase transition now
we study the kinetics of the phase separation in detail. As an example
we consider the nucleation process in a supersaturated vapour with a
finite system size and a limited overall particle number.
Several models have been developed to describe this phase transition.
An overview was already given in Chapter 1. We divide the theories now
into two groups, deterministic and statistic theories.
A deterministic description of the phase transition is appropriate to
investigate the time dependent evolution of the system when overcritical clusters have been already established. Because fluctuations are
neglected for a deterministic description the formation of critical
clusters cannot be explained. Therefore,usually a source term will be
added to the deterministic equation which generates the critical clusters. A possible form of this term is given by the classical nucleation rate. Deterministic equations to describe the growth of overcritical clusters are discussed in Chapter 4.5. as well as in the Chapters
5 and 6.
A more proper consideration of the whole process of nucleation and
growth of clusters is held from a statistic point of view. The formation of small clusters has to describe as a microscopic process (EBE-
LING et al. 1976). But from such an investigation a lot of problems
arise in the range of larger clusters because of the very complicated
elementary processes.
We propose therefore a stochastic description for the formation of
clusters by free particles and their growth up to a critical size. The
existence of a nucleation barrier is known from the thermodynamic investigation. The cross-over of this barrier is an intrinsic stochastic
process which can be explained only by the consideration of the fluctuations in the system. In a deterministic sense undercritical clusters have to diminish again.
The stochastic description restricts itself to a mesoscopic time scale.
That means a scale where the microscopic processes are not considered
in detail but reflected by small changes of the macroscopic parameters
of the system (e.g. pressure, temperature).

An advantage of the stochastic description is its close relation to
computer simulation experiments to justify the theory. It is also easy
to derive deterministic equations from a stochastic description as it

will be shown in Chapter 4.5.

The model we introduce in the following to derive a stochastic theory of nucleation and growth of clusters in a supersaturated vapour is something different but nevertheless in accordance to the model used for the thermodynamic description. Again, we consider a closed and finite system with a fixed system volume V and a fixed overall particle number N in the gaseous state:

$$N = const., \quad V = const. \tag{4.1}$$

As it is known from experiments in general a carrier gas is used to transport the latent heat which will be released during the condensation process. This carrier gas is uncondensable for the given constraints. The overall particle number therefore is divided into the particle number of the carrier gas (N_0) being constant and the particle number of the condensable vapour (N_m) being constant too.

$$N = N_0 + N_m \tag{4.2}$$

Due to interactions between the particles of the condensable supersaturated vapour a number of particles is bound in clusters and a distribution of clusters and free particles in the gas exists:

$$\underline{N} = \left\{ N_1 N_2 N_3 \ldots N_{n-1} \; N_n \; N_{n+1} \ldots N_N \right\} \tag{4.3}$$

$\underline{N}$ denotes the discrete cluster distribution where N_1 is the number of the free particles of the condensable vapour (monomers), N_2 the number of bound states of two particles (dimers) and so on.

Because of the limited number of particles it holds:

$$N_m = \sum_{n=1}^{N} n \, N_n = const. \tag{4.4}$$

n is the number of particles bound in the cluster, the number of clusters consisting of n particles is denoted by N_n.

For the maximum number of clusters it follows from (4.4)

$$0 \leq N_n \leq N_m/n \; . \tag{4.5}$$

We can now discuss the two limiting cases:

(i) For $N_0 = 0$ we have isoenergetic conditions. The inner energy of the system should be constant, because no condensation heat is transported to the background by means of the carrier gas.

(ii) If $N_0 \gg N_m$, the latent heat of condensation will be completely brought back in a short time and the temperature of the gaseous system is held constant.

While the first case (isoenergetic nucleation) is discussed in subsequent papers (EBELING et al., 1988; BUDDE, MAHNKE, 1987) we restrict ourselves here to the second case of the isothermal nucleation process. The influence of the carrier gas on the phase transition is not considered furthermore. So we fix the thermodynamic constraints now as follows:

$$N = \text{const.}, \quad V = \text{const.}, \quad T = \text{const.} \tag{4.6}$$

with N being the condensable particle number. The thermodynamic constraints are chosen in such a way that the pressure of a supposed perfect vapour consisting of free particles only is larger than the equilibrium pressure $p'(T)$ of the saturated vapour at the same temperature:

$$p = Nk_BT/V > p'(T) \tag{4.7}$$

This allows us to introduce again the initial supersaturation y_0 (eq. (3.7)) used already in Chapter 3: $y_0 = p/p'(T)$.

We now derive the canonical partition function for the system with the assumed cluster distribution $\underline{N}$ (eq. (4.3)). In general $Z(T,V,\underline{N})$ can be calculated by (EBELING et al., 1976)

$$Z(T,V,\underline{N}) = \int\limits_{C(\underline{N})} \exp\left\{ -\frac{1}{k_BT} H(q_1\ldots q_N p_1\ldots p_N) \right\} dq_1\ldots dq_N dp_1\ldots dp_N \tag{4.8}$$

$H(q_1\ldots p_N)$ is the Hamiltonian of the N particles of the system with full interactions, where q_k are the space coordinates and p_k are the momentum coordinates of the particles ($k=1,\ldots,N$). The integration is carried out for the subspace

$$C(\underline{N}) = C(N_1\ldots N_N)$$

of the assumed particle configuration $\underline{N}$.

Supposing an ideal mixture of clusters and free particles we get from eq. (4.8) with the correct normalization (BECKER, 1961)

$$Z(T,V,\underline{N}) = \prod_{n=1}^{N} \frac{1}{h^{3N_n} N_n!} \int \exp\left\{ -\frac{1}{k_BT} H_n \right\} dq_1\ldots dq_{N_n} dp_1\ldots dp_{N_n} \tag{4.9}$$

q_{N_n} and p_{N_n} are now the space coordinates and momentum coordinates of the clusters of size n. H_n is the Hamiltonian of the cluster with the particle number n and the mass m_n:

$$H_n = \frac{p_n^2}{2m_n} + f_n \tag{4.10}$$

f_n denotes a potential contribution which depends only on the size n of the cluster. Because of the ideal mixture no contributions due to interactions between clusters and monomers arise.

The free energy of the assumed cluster distribution is obtained as follows:

$$F(T,V,N_1 \ldots N_N) = - k_B T \ln Z(T,V,N_1 \ldots N_N) \qquad (4.11)$$

After integration of eq. (4.9) with respect to eq. (4.10) we get for the free energy (SCHIMANSKY-GEIER et al., 1986; SCHWEITZER, 1986a)

$$F(T,V,N_1 \ldots N_N) = \sum_{n=1}^{N} N_n \left\{ f_n + k_B T (\ln \frac{N_n}{V} \lambda_n^3 - 1) \right\} \qquad (4.12)$$

$\lambda_n = h/(2 \pi m_n k_B T)^{1/2}$ is the de Broglie wave length. To determine the potential term f_n we choose a first approximation similar to the theory of atomic nuclei which includes only volume and surface effects:

$$f_n = - A \, n + B \, n^{2/3} \qquad (4.13)$$

The first term of eq. (4.13) corresponds to the binding energy in the cluster, but the second term to the surface energy.

The constant A means the binding energy of a particle in the cluster. It can be estimated by the molar evaporation heat H_v:

$$A = H_v / N_A$$

Another but quite equal expression for A was derived in comparison with thermodynamic results (SCHWEITZER and SCHIMANSKY-GEIER, 1987):

$$A = - k_B T \ln \frac{n'(T)}{k_B T} \lambda_1^3 \qquad (4.14)$$

The surface energy is proportional to the surface area and the surface tension σ. Assuming a spherical cluster it yields for the constant B:

$$B = 4 \pi \left(\frac{4 \pi}{3} c_\alpha \right)^{-2/3} \sigma \qquad (4.15)$$

c_α is the particle density in the cluster ($c_\alpha = \varrho_\alpha N_A$, N_A = Avogadro constant). Due to the classical droplet model the surface tension σ and the particle density c_α are assumed to be constant.

Fig. 4.1 illustrates the potential term f_n (eq. (4.13)) by presentation of the derivation $\partial f_n / \partial n$ versus cluster size n.

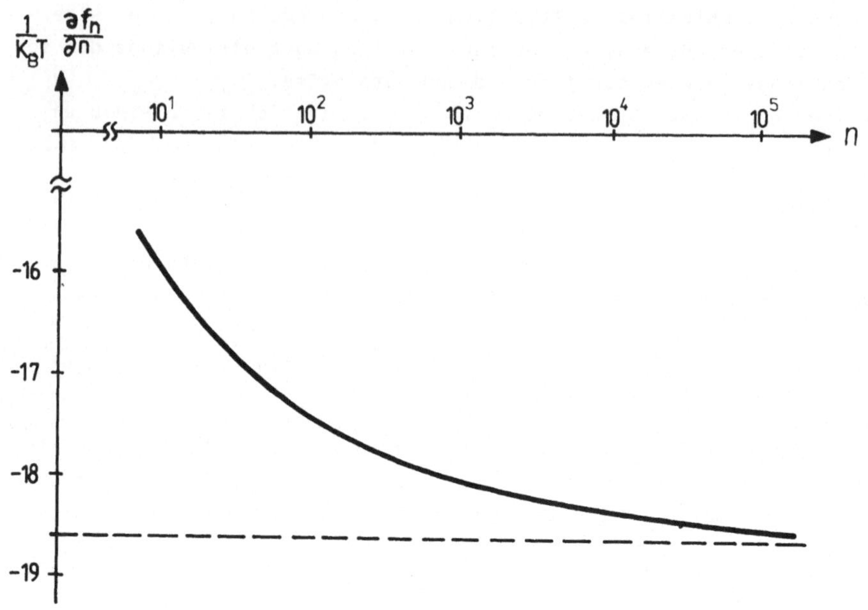

Fig. 4.1
Derivation $\partial f_n / \partial n$ of the potential term f_n (eq. (4.13)) versus cluster size n in the range of large clusters. The calculation was done
for water at T = 293 K. It yields for the constants A = 18.639 $k_B T$
(eq. (4.14)) and B = 8.041 $k_B T$ (eq. (4.15))

For very large clusters the plot converges into the value for the
constant A meaning the binding energy per particle in a macoscopic
planar liquid phase.
The ansatz (4.13) for f_n is valid only for large clusters, where a
real surface can be divided from the inner part of the cluster. For
small clusters (compare also Fig. 4.2)we propose another form of f_n
(SCHWEITZER,1986a):

$$f_n = - \frac{a}{2} n (n - 1) \tag{4.16}$$

Thus we find for f_n in the range of small values of n:

$\quad f_1 = 0 \qquad$ for monomers
$\quad f_2 = -a \qquad$ dimer binding energy
$\quad f_3 = - 3a \tag{4.17}$

and so on. The dimer binding energy has to be known from experiments.

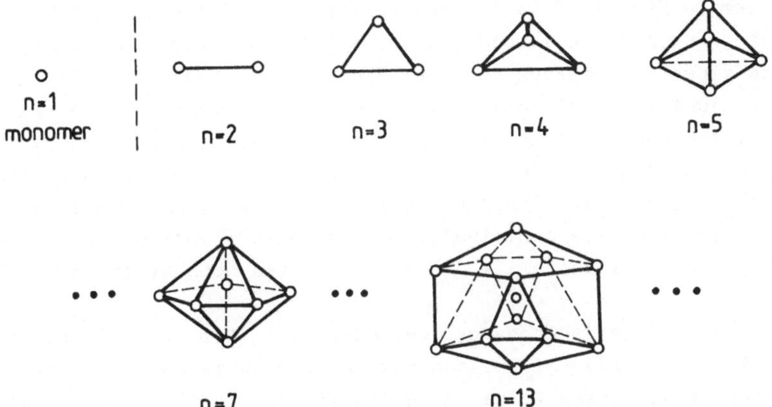

Fig. 4.2
Sketch of small clusters consisting of view particles only. The sketch
illustrates the number of bonds in the cluster. (Adapted from JORTNER,
1984)

In this context the investigations concerning the potential term f_n
for small and large water clusters given by MAHNKE and BUDDE (1987,
1988) should be mentioned. A Padé approximation by EBELING (1981) gives
a possible interpolation of the expressions for $\partial f_n / \partial n$ in the range
of small and large clusters.

The free energy (eq. (4.12)) includes the contributions of the pressure p and the Gibbs free energy; F = G - pV. For the pressure it follows:

$$p = - \frac{\partial F}{\partial V} = \frac{k_B T}{V} \sum_{n=1}^{N} N_n \qquad (4.18)$$

while the Gibbs potential G is given by:

$$G = \sum_{n=1}^{N} \mu_n N_n \qquad (4.19)$$

with

$$\mu_n = \frac{\partial F}{\partial N_n} = f_n + k_B T \ln \frac{N_n}{V} \lambda_n^3 \qquad (4.20)$$

being the chemical potential of a cluster of size n. Detailled investigations of the free energy (eq. (4.12)) are given by SCHWEITZER and
TIETZE (1987).
The free energy of the ideal gas mixture of clusters and free particles (eq.(4.12)) has got a something different form compared with the

expressions from thermodynamics. For the discussion of the heterogeneous system in Chapter 3 we introduced the idea of the two phases:
the vapour phase and the liquid phase spatially distributed in drops.
Here we consider one gaseous phase with different kinds of sorts, given by the clusters of different sizes n. Consequently the pressure
is given by the sum of the partial pressures of all sorts of clusters.
While in Chapter 3 μ_α was the chemical potential related to one particle of the liquid phase, here μ_n is the chemical potential of the
cluster. It includes the contribution of the mixing entropy. The surface energy of the cluster $F_o = \sigma A$ is represented here by the term
$B \cdot n^{2/3}$ in f_n.

The expression of the free energy used here is a more adapted one to
a stochastic description of the nucleation process including statistical considerations. There arise no contradictions to the thermodynamic
investigations of Chapter 3.

4.2. Kinetic Assumptions and Master Equation

The nucleation process means the formation of clusters and their growth
and shrinkage. The cluster evolution is represented by the time depen-
dent development of the distribution $\underline{N} = \{N_1, N_2, \ldots, N_n, \ldots N_N\}$ (eq.
(4.3)). To discuss this evolution we suppose the following assumptions:
(i) The growth and shrinkage of a cluster is due only to an attachment
    or evaporation of monomers (Fig. 4.3).

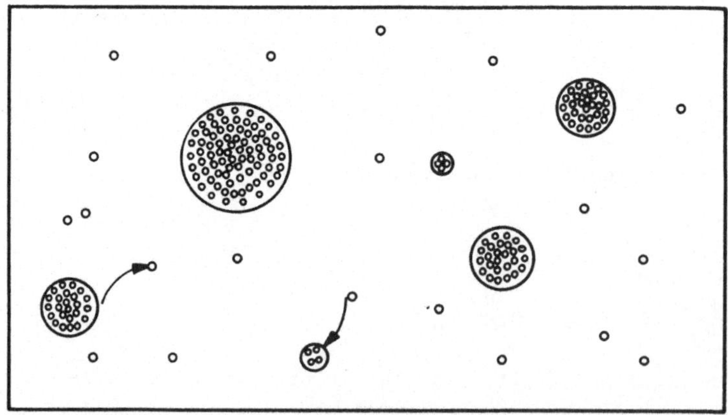

Fig. 4.3: Typical cluster distribution $\underline{N}$ consisting of particles of
different sizes n. The elementary processes of attachment and evaporation of monomers are shown by arrows.

In terms of chemical kinetics this process can be represented by
the reaction (compare also eq. (1.b)):

$$A_n + A_1 \underset{w^-}{\overset{w^+}{\rightleftharpoons}} A_{n+1} \tag{4.21}$$

w^+ and w^- are the probabilities per time unit for the reaction in
the given direction. They will be specified afterwards.

(ii) Interactions between clusters with a size $n \geqslant 2$, like coagulation
     processes or collisions between two or more clusters are not ta-
     ken into account. Also a break of a cluster into pieces is not
     considered. The probabilities of these events should be neglec-
     tible in comparison with the probabilities of the reactions
     (4.21).

Due to the interaction assumed in eq. (4.21) every possible configuration $\underline{N}$ of particles in clusters is found only with a certain probability. We define

$$P(\underline{N},t) = P(N_1,N_2,N_3,\ldots,N_n\ldots N_N,t) \tag{4.22}$$

to be the probability that a certain cluster distribution $\underline{N}$ exists at
the time t.

The stochastic description of the nucleation process deals with the
investigation of the time dependence of $P(\underline{N},t)$. The change of $P(\underline{N},t)$
with time can be described by a Master equation (for the concept of
Master equations see e.g. VAN KAMPEN, 1970; GARDINER, 1984; HAKEN,
1978; RÖPKE, 1987). Because $\underline{N}$ is a vector of discrete state the Master
equation has got the form:

$$\frac{\partial P(\underline{N},t)}{\partial t} = \sum_{\underline{N}' \neq \underline{N}} \left\{ w(\underline{N}|\underline{N}')P(\underline{N}',t) - w(\underline{N}'|\underline{N})P(\underline{N},t) \right\} \tag{4.23}$$

The quantities $w(\underline{N}'|\underline{N})$ are the transition probabilities per unit time
for the transition from $\underline{N}$ to $\underline{N}'$. $\underline{N}'$ specifies those distributions
which are attainable from the assumed distribution $\underline{N}$ via the reaction
(eq. (4.21)).

Eq. (4.23) allows us to introduce a probability flux J

$$J(\underline{N}|\underline{N}') = w(\underline{N}|\underline{N}')P(\underline{N}',t) - w(\underline{N}'|\underline{N})P(\underline{N},t) \tag{4.24}$$

J is the stochastic pendant to the netto rate discussed in the classical nucleation theory. Substitution of eq. (4.24) into the master
equation results in:

$$\frac{\partial P(\underline{N},t)}{\partial t} = \sum_{\underline{N}'} J(\underline{N}|\underline{N}') \tag{4.25}$$

The stationary solution of the master equation requires that the probability to find a given cluster distribution does not change with time. From the condition $\partial P(\underline{N},t)/\partial t = 0$ we find

$$\sum_{\underline{N}'} J(\underline{N}|\underline{N}') = 0 \tag{4.26}$$

This is the so called Kirchhoff solution (HAKEN, 1978). For finite systems this condition is only fulfilled if it yields:

$$J(\underline{N}|\underline{N}') = 0 \tag{4.27}$$

This more restrictive form of eq. (4.27) is the so called condition of detailed balance. It gives the equilibrium condition for finite systems (LANDAU and LIFSHITZ, 1983).
The condition (4.27) means:

$$w(\underline{N}|\underline{N}')P^O(\underline{N}') = w(\underline{N}'|\underline{N})P^O(\underline{N}) \tag{4.28}$$

where $P^O(\underline{N})$ is the probability that $\underline{N}$ is the equilibrium cluster distribution.
$P^O(\underline{N})$ can be derived from microscopic considerations (SCHIMANSKY-GEIER et al. 1986). Let $u(q_1 \ldots q_N p_1 \ldots p_N)$ be the probability distribution for the N particles in the phase space of the space and momentum coordinates of all particles. In thermodynamic equilibrium the following relation is valid (BECKER, 1961):

$$u^O(q_1 \ldots p_N) = \frac{1}{Z(T,V,N)} \exp\left\{- \frac{1}{k_B T} H(q_1 \ldots q_N p_1 \ldots p_N)\right\} \tag{4.29}$$

$H(q_1 \ldots p_N)$ is the Hamiltonian of the N particles as used already in eq. (4.8). $Z(T,V,N)$ is the canonical partition function of the N particle system in an atomic picture. For the given thermodynamic constraints $Z(T,V,N)$ is a constant and gives the normalization. We now define the probability of the equilibrium distribution $P^O(\underline{N})$ by an integral of $u^O(q_1 \ldots p_N)$ over the subspace $C(\underline{N})$:

$$P^O(\underline{N}) = \int_{C(\underline{N})} u^O(q_1 \ldots p_N)\, dq_1 \ldots dp_N \tag{4.30}$$

With eq.(4.29) it yields:

$$P^O(\underline{N}) = \frac{1}{Z(T,V,N)} \int_{C(\underline{N})} \exp\left\{- \frac{1}{k_B T} H(q_1 \ldots p_N)\right\} dq_1 \ldots dp_N \tag{4.31}$$

Use of eq. (4.8) allows us to introduce the canonical partition function for the assumed cluster distribution $Z(T,V,\underline{N})$ into eq. (4.31). We receive finally with respect to eq. (4.11) the probability of the

94

equilibrium distribution in the form:

$$P^0(\underline{N}) = \frac{1}{Z(T,V,N)} \exp\left\{-\frac{1}{k_B T} F(T,V,\underline{N})\right\} \qquad (4.32)$$

Eq. (4.32) means that the probability of the equilibrium distribution has its maximum just in the case when the free energy has its minimum. The equilibrium cluster distribution therefore can be found from the extremum condition:

$$\frac{\partial P(N)}{\partial N_n} = 0, \text{ respectively } \left.\frac{\partial F(T,V,N)}{\partial N_n}\right|_{T,V,N} = 0 \qquad (4.33)$$

We find

$$\frac{N_n^0}{V} \lambda_n^3 = \left(\frac{N_1^0}{V} \lambda_1^3\right)^n \exp\left\{-\frac{f_n}{k_B T}\right\} \qquad n=2,\ldots,N \qquad (4.34)$$

where $N_1^0 = N - \sum\limits_{n=2}^{N} nN_n^0$.

Due to this boundary condition eq. (4.34) means a system of equations which has to be solved simultaneously. The equilibrium solution can be written in terms of a mass action law (MAHNKE, BUDDE, 1988):

$$\frac{N_n^0}{(N_1^0)^n} = C_n(T,V) = \left(\frac{\lambda_1^3}{V}\right)^{n-1} n^{3/2} \exp\left\{-\frac{f_n}{k_B T}\right\} \qquad (4.35)$$

$C_n(T,V)$ is the mass action constant. Using the denotation of the chemical potential μ_n (eq. (4.20)) the mass action law has the known form (EBELING et al., 1976):

$$\mu_n^0 = n \, \mu_1^0 \qquad (4.36)$$

For a further discussion of the stochastic evolution of the phase transition we need now proper transition probabilities to describe the attachment or evaporation of monomers to/from clusters. Due to the condition of detailed balance we only need one kinetic assumption with respect to the reaction (4.21). The transition probability for the opposite process can be derived by means of the free energy as follows:

$$w(\underline{N}|\underline{N}') = w(\underline{N}'|\underline{N})\exp\left\{\frac{1}{k_B T} (F(T,V,\underline{N}')-F(T,V,\underline{N}))\right\} \qquad (4.37)$$

In agreement with earlier investigations (EBELING, 1981) we assume that the probability of the attachment of a monomer to a cluster of size n increases with the surface of the cluster and with the density of free particles and the number of clusters of size n. In this way we get for the special process of attachment:

$$w(\underline{N}'|\underline{N}) = w(N_1-1\ldots N_n-1\ N_{n+1}+1\ldots N_N|N_1\ldots N_n N_{n+1}\ldots N_N)$$

$$\equiv w_n^+(N_1 N_n) = \alpha n^{2/3} N_n\ N_1/V\ ;$$

$$N_1 = N - \sum_{n=2}^{N} n N_n \qquad\qquad (4.38)$$

Eq. (4.38) demonstrates in terms of the argument of $w(\underline{N}'|\underline{N})$ the change of the cluster distribution $\underline{N}$ due to the attachment of one monomer to the cluster of size n. α is a constant which scales the time. In this Chapter α is set equal to one. In general the parameter α has to consider the specific properties of the surface, like surface tension, composition of the surface and others. It includes also the sticking coefficient measuring the rate of monomers which stick at the surface compared with the total number of free particles which collide with the cluster per unit time. From experiments a value of the sticking coefficient of about 0.04 is known (KOTAKE and GLASS, 1978) but the results differ strongly.

We note that in the case of creating a dimer the transition probability differs something. In this case it reads:

$$w(N_1-2\ N_2+1\ldots N_N|N_1 N_2\ldots N_N)$$

$$\equiv w_1^+(N_1) = \alpha \frac{N_1(N_1-1)}{V} \qquad\qquad (4.39)$$

We would like to underline that the probabilities of the growth of clusters of different sizes are correlated by the vapour pressure because the number of particles is conserved. Therefore in finite systems the stochastic cluster evolution cannot longer be described in terms of a one-dimension random walk process (cf. SHUGARD and REISS, 1976; GILLESPIE, 1981).

The transition probability per unit time for the evaporation of one particle from a cluster of size n is received from the condition of detailed balance (eq. (4.37)) using the free energy $F(T,V,\underline{N})$ (eq. (4.12)). We find this transition probability in the form:

$$w(\underline{N}'|\underline{N}) = w(N_1+1\ldots N_{n-1}+1\ N_n-1\ldots N_N|N_1\ldots N_{n-1} N_n\ldots N_N)$$

$$\equiv w_n^-(N_n) = \alpha n^{2/3}\ N_n\ \left(\frac{n-1}{n}\right)\ 13/6\ \frac{1}{\lambda_1^3}\ \exp\left\{\frac{1}{k_B T}\ (f_n-f_{n-1})\right\} \qquad (4.40)$$

The argument of $w(\underline{N}'|\underline{N})$ gives the change of the cluster distribution $\underline{N}$ due to the evaporation of a monomer from the cluster of size n. This transition probability is proportional to the surface area and to the number of clusters of size n again. The special properties of the surface are reflected in the parameter α .

Neglecting the term $(n-1/n)^{13/6}$ which is nearly equal to one and using the ansatz for f_n (eq. (4.13)) valid for large clusters the transition probability (eq. (4.40)) can be written in the form:

$$w_n^-(N_n) = \alpha\, n^{2/3}\, N_n\, \frac{p'}{k_B T}\, \exp\left\{\ \frac{2}{3}\frac{B}{k_B T}\, n^{-1/3}\right\} \tag{4.41}$$

Let us note, that the transition probability of evaporation in our model is not determined by the whole cluster distribution as the transition probability of attachment. It depends only on the conditions of the cluster itself.

For the model presented here the transition probabilities of all processes different from the reaction (4.21) are assumed to be zero. But we remark that the theory can be expanded without serious diffuculties to consider also the mentioned above events, if the transition probabilities for the collisions of clusters are chosen in an analogous form of eq.(4.38). The master equation (4.23) reads in an explizit form with the introduced transition probabilities (SCHIMANSKY-GEIER et al., 1986)

$$
\begin{aligned}
\frac{\partial}{\partial t}\, P(N_1\ldots N_N) =\ & w_2^-(N_2+1)P(N_1-2\ N_2+1\ldots N_N,t)\\
& -\ w_1^+(N_1)P(N_1\ldots N_N,t)\\
& +\ w_3^-(N_3+1)P(N_1-1\ N_2-1\ N_3+1\ldots N_N,t)\\
& +\ w_1^+(N_1+2)P(N_1+2\ N_2-1\ N_3\ldots N_N,t)\\
& -\ [w^+{}_2(N_1 N_2)+w_2^-(N_2)]\ P(N_1\ldots N_N,t)\\
& +\ \sum_{n=3}^{N}\left\{w_{n+1}^-(N_{n+1}+1)P(N_1-1\ldots N_n-1\ N_{n+1}+1\ldots N_N,t)\right.\\
& +\ w_{n-1}^+(N_1+1\ N_{n-1}+1)P(N_1+1\ldots N_{n-1}+1\ N_n-1\ldots N_N,t)\\
& \left. -\ [w_n^+(N_1 N_n)+w_n^-(N_n)]\ P(N_1\ldots N_n,t)\right\} \tag{4.42}
\end{aligned}
$$

The complicated form of eq. (4.42) is caused by the numerous possibilities to change the considered cluster distribution $\underline{N}$ via reactions of the type (eq. (4.21)). The dimension of the stochastic process is given by the number of different sorts of clusters in the system.

4.3. Results of Computer Simulations

4.3.1. Stochastic Dynamics Technique

The master equation (4.42) has to be solved by use of computer simulation techniques. The most discussed techniques are Monte Carlo Simulations (for an overview see e.g. BINDER 1979, 1984) and molecular dynamic simulations (e.g. ZUREK and SCHIEVE, 1981). In the fields of phase transitions, clustering and nucleation, description of small clusters and effects on the liquid-vapour surface there exists a variety of interesting results (ABRAHAM, 1981, 1982; BINDER, 1977b; KALOS et al., 1978; PENROSE et al., 1978, 1983, 1984; HALE and WARD, 1982; HEERMANN et al., 1984; RAO et al., 1978; KOCH and LIEBMANN, 1983). The investigations have been carried out mainly for 2d- or 3d-Ising systems or Lenard-Jones systems.

But the establishment of a cluster configuration in a metastable finite system and its time dependent evolution to the final state is not considered in detail before. We devote this Chapter therefore to the investigation of this problem. To solve the master equation (4.42) we use the stochastic simulation technique (see also FEISTEL and EBELING, 1978; EBELING and SCHIMANSKY-GEIER, 1979; FEISTEL, 1979; MALCHOW and SCHIMANSKY-GEIER, 1985; GILLESPIE, 1979, 1980). The basic ideas of the stochastic dynamics and its close relation to the stochastic description given above are explained in the following.

We assume that at a given time t_o a certain cluster distribution $\underline{N}$ exists, that means $P(\underline{N},t_o)$ is equal to one. The probabilities of all other distributions $\underline{N}'$ at the same time are equal to zero. The master equation (4.23) is then reduced to:

$$\frac{\partial P(N,t)}{\partial t} = - \sum_{\underline{N}' \neq \underline{N}} w(\underline{N}'|\underline{N})P(\underline{N},t), \qquad t \geq t_o \qquad (4.43)$$

The probability $P(\underline{N},t)$ to find the distribution $\underline{N}$ decreases with an increasing time. The solution of eq. (4.43) is given by (BHARUCHA-REID, 1969):

$$P(\underline{N},t) \sim \exp\left\{ - \frac{t}{\vartheta_N} \right\} \qquad (4.44)$$

ϑ_N signifies the mean life time of the cluster distribution $\underline{N}$. For $t \ll \vartheta_N$ $P(\underline{N},t)$ is nearly equal to one, while for $t \gg \vartheta_N$ $P(\underline{N},t)$ tends to zero. The mean life time is given by all possibilities to leave the state $\underline{N}$. It yields:

$$\vartheta_N = 1/ \sum_{\underline{N}' \neq \underline{N}} w(\underline{N}'|\underline{N}) \qquad (4.45)$$

The real life time $t_{\underline{N}}$ of the configuration $\underline{N}$ is a randomly distributed
value. From eq. (4.44) we find the ansatz:

$$t_{\underline{N}} = - \vartheta_{\underline{N}} \ln \text{RNDM}(0,1) \tag{4.46}$$

RNDM(0,1) is an equal distributed random number between zero and one
(but zero itselves has to eliminate). For $t = t_{\underline{N}}$ the distribution $\underline{N}$
changes into $\underline{N}'$ with the probability

$$\frac{w(\underline{N}'|\underline{N})}{\sum\limits_{\underline{N}'} w(\underline{N}'|\underline{N})} : \underline{N} \rightarrow \underline{N}' \tag{4.47}$$

There exists a big number of possibilities to change $\underline{N}$ in detail which
is given by the number of possible transition probabilities. To sepa-
rate one of these processes a second random number $\text{RNDM}(0, \sum\limits_{\underline{N}'} w(\underline{N}'|\underline{N})$
is used. The sketch gives the sum over all transition probabilities:

$$+ \cdots \overline{\quad \overset{w^+_n}{\quad} \overset{w^-_n}{\quad} \overset{w^+_{n+1}}{\quad} \overset{w^-_{n+1}}{\quad} \quad} \cdots$$

$$0 \qquad\qquad\qquad\qquad\qquad\qquad\qquad \sum\limits_{\underline{N}'} w(\underline{N}'|\underline{N})$$

$$\text{RNDM}(0, \sum\limits_{\underline{N}'} w(\underline{N}'|\underline{N}))$$

The random number is located in a certain interval of the possible
transition probabilities. In the given example w^-_n is randomly chosed.
That means a cluster of size n evaporates a monomer and gets the size
n-1. The cluster distribution in this special case changes as follows:

$$(N_1+1 \ N_2 \ldots N_{n+1}+1 \ N_n-1 \ldots N_N \mid N_1 N_2 \ldots N_{n-1} N_n \ldots N_N)$$

For the new particle configuration a new mean life time is calculated
by means of the new possible transition probabilities. For given transition probabilities in this way the whole process of formation of
clusters and the time dependent evolution of the cluster distribution
can be stimulated. The clock of the system is given by the sum over all
real life times of the different configurations which have been established.

4.3.2. Evolution of a Single Cluster

To demonstrate the stochastic process of nucleation first a simple
case is discussed (see also SCHWEITZER, 1987). We assume a system where
only one cluster evolves. This cluster can be surrounded only by free
particles or additionally by a distribution of other clusters kept constant. An ansatz to reduce the development of a cluster distribution
to the case of one evolving cluster is proposed by SCHIMANSKY-GEIER
et al. (1986) (for finite systems) and GILLESPIE (1981) (for infinite
systems). If only a single cluster can be formed in the finite system
it holds for the overall particle number

$$N = N_1 + n = \text{const.} \qquad (4.48)$$

with N_1 being the number of monomers and n being the cluster size.
The probability to find the cluster with size n at a given time t is
denoted by P(n,t). The master equation (4.23) reads in this case:

$$\frac{\partial P(n,t)}{\partial t} = \sum_{n' \neq n} \left\{ w(n|n')P(n',t) - w(n'|n)P(n,t) \right\} \qquad (4.49)$$

Because the growth and shrinkage of the single cluster occurs by the
attachment or evaporation of free particles, the states n' can be specified as (n+1) and (n-1), respectively:

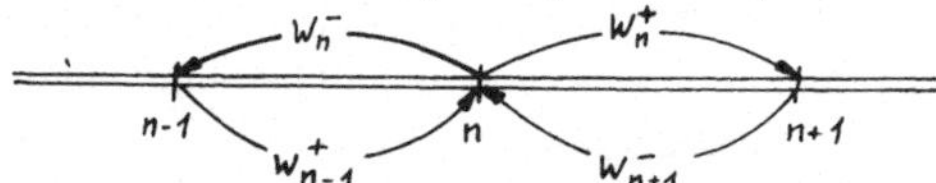

The master equation (4.49) is written then in the form:

$$\frac{\partial P(n,t)}{\partial t} = w^+_{n-1}P(n-1,t) + w^-_{n+1}P(n+1,t)$$
$$- \left[w^+_n + w^-_n \right] P(n,t) \qquad (4.50)$$

Eq. (4.50) describes a typical one-dimensional random walk process.
   We can distinguish betweeen two characteristic terms: the gain term
given by the probability flux from the states (n+1) and (n-1) to the
state n and the lost term of the probability flux from state n to the
other states.
The transition probabilities in the considered case are only functions
of the cluster size n because of eq. (4.48). In accordance with the
eqs. (4.38) and (4.40) they have got the form:

$$w(n+1|n) \equiv w^+_n = \alpha' \; n^{2/3} \; \frac{(N-n)k_B T}{p'V}$$

$$w(n-1|n) \equiv w^-_n = \alpha' \; n^{2/3} \; \exp\left\{\frac{2}{3} \frac{B}{k_B T} \; n^{-1/3}\right\} \qquad (4.51)$$

We have introduced here a different form of the parameter α: It is α' = $\alpha \frac{p'}{k_B T}$. The term p_1 = $(N-n)k_B T/V$ gives the recent partial pressure of the free particles and $y = p_1/p'$ is the recent supersaturation of the system. Instead of this the initial supersaturation was given by y_0 (eqs. (3.7) and (4.5)). For the finite system the transition probabilities w_n^+ and w_n^- are presented in Fig. 4.4. In dependence on the initial supersaturation y_0 we find either two points of intersection or one or no intersection. These points of intersection give the possible equilibrium states of the cluster in agreement with the thermodynamic investigations of Chapter 3. The point of a lower value of n corresponds to the critical state, while the intersection for the larger value of n gives the stable cluster size. It is to be seen that in the range of small cluster sizes (for $n < n_{cr}$) the transition probability for the evaporation of a formed cluster is larger than the probability of attachment. That means from a deterministic point of view it is not able to explain the formation of the critical cluster. This process is intrinsic stochastically due to the influence of fluctuations. The overcritical cluster grows until reaching its stable state, because it yields $w_n^+ > w_n^-$ (for $n_{cr} < n < n_{st}$). But a cluster with a size larger than the stable size has to shrink again.
The figure shows also the dependence of the critical and the stable cluster size on the initial supersaturation. This fact was discussed in detail already in Chapter 3. The critical supersaturation when both equilibrium states coincides, is given by eq. (3.35). For a comparison with the finite system Fig. 4.5 gives the transition probabilities for the infinite system. In this case the partial pressure of the free particles is always constant, and the recent supersaturation is equal to the initial supersaturation. Consoquently an overcritical cluster has to grow further and no stable state exists.

The stochastic evolution of the cluster can be simulated by means of the stochastic dynamics derived before. The results are presented in Fig. 4.6 for the finite system. We denote a significant dependence of the cluster size evolution on the initial supersaturation y_0. For slightly larger values of y_0 the stable state is reached much more faster and will not be leaved. But for small initial supersaturations it takes the cluster much more longer to occupy its thermodynamically stable state and the time of staying in this state is very short.

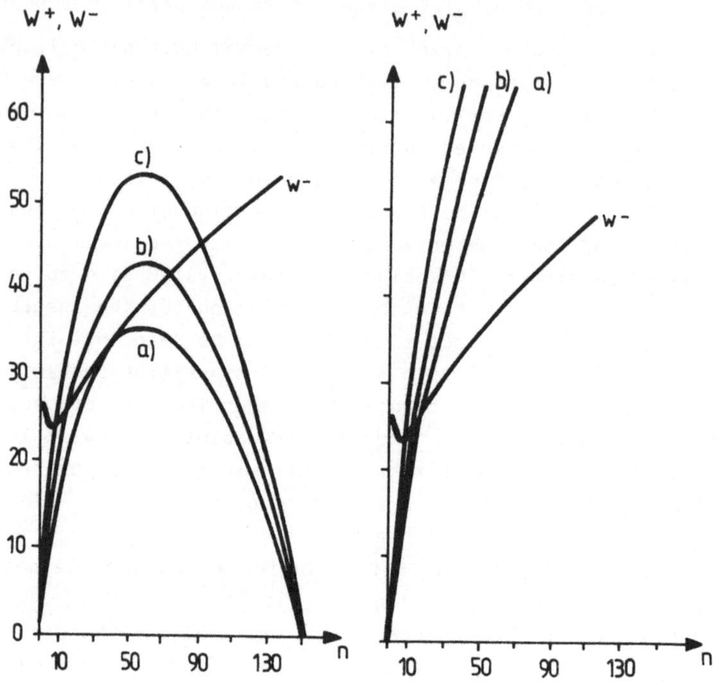

Fig. 4.4:   Fig. 4.5:
Transition probabilities w_n^+ and w_n^- for the finite system (eq. (4.51))
(Fig. 4.4) and for the infinite system (Fig. 4.5). Initial supersa-
turation (a) y_o = 3.86; (b) y_o = 4.63; (c) y_o = 5.79; overall particle
number N = 150 (for Fig. 4.4)
(From: SCHWEITZER, 1987)

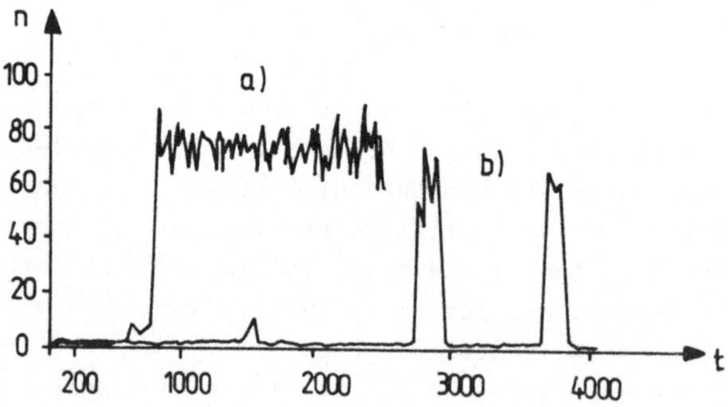

Fig. 4.6:
Stochastic evolution of the cluster size n versus time. Initial su-
persaturation (a) y_o = 4.63; (b) y_o = 4.21; overall particle number
N = 150
(From: SCHWEITZER, 1987)

For the infinite system a stationary solution $P^S(n)$ of the master
equation (eq. (4.50)) can be obtained, where the probability flux
does not vanish. From the condition $\partial P(n,t)/\partial t = 0$ we find:

$$J(n|n+1) = w_n^+ P^S(n) - w_{n+1}^- P^S(n+1) = J^S = \text{const.} \qquad (4.52)$$

$$\text{for } n = 1,\ldots,N-1$$

J^S is the stationary probability flux being constant. Eq. (4.52) means
a recursive system of equations. The stationary probability $P^S(n)$ can
be derived from eq.(4.52) in the form (ULBRICHT et al., 1986):

$$P^S(n) = P^S(m) \prod_{j=m+1}^{n} \frac{w_{j-1}^+}{w_j^-} - J^S \left(\sum_{k=m}^{n-1} Q_k \right) ; \quad m < n \qquad (4.53)$$

with

$$Q_k = \frac{1}{w_k^+} \prod_{j=m+1}^{k} \frac{w_{j-1}^+}{w_j^-}$$

For a known stationary probability of the cluster in state m (m   n)
and a known stationary probability flux $P^S(n)$ can be calculated easy
with given transition probabilities. On the other side the stationary
probability flux can be expressed in the form (SCHWEITZER et al., 1987)

$$J^S = \frac{\phi_n - \phi_m}{\sum_{k=m}^{n-1} R_k} \qquad (4.54)$$

with

$$R_k = \left[w_k^+ \prod_{j=2}^{k} \frac{w_{j-1}^+}{w_j^-} \right]^{-1} ; \quad \phi_i = P^S(i) \left[\prod_{j=2}^{i} \frac{w_{j-1}^+}{w_j^-} \right]^{-1} \qquad i=m,n$$

Eq.(4.54) can be interpreted in comparison with the theory of clootric
currents (BECKER and DÖRING, 1935): J^S means in this case a constant
current which flows over a number of resistors R_k connected in file.
ϕ_m and ϕ_n give the "tensions" in the states m and n. The difference
has to be positive to obtain a current J^S.
If we choose m = 1 and n = k to be an absorber state, that means $P^S(k)$
vanishes, eq. (4.54) results in:

$$J^S = P^S(1) \left\{ \sum_{n=1}^{k-1} w_n^+ \prod_{j=2}^{n} \frac{w_{j-1}^+}{w_j^-} \right\}^{-1} ; \quad P^S(k) = 0 \qquad (4.55)$$

In the given form J^S represents a stochastic pendant to the classical
steady-state nucleation rate discussed in Chapter 1.
With the introduction of a stationary potential Ω_n (GRAHAM, 1973):

$$\mathfrak{R}_n = - k_B T \sum_{j=2}^{n} \ln \frac{w_{j-1}^+}{w_j^-} \tag{4.56}$$

the stationary probability flux finally is obtained as follows:

$$J^S = P^S(1) \left\{ \sum_{n=1}^{k-1} w_n^+ \exp(- \frac{\mathfrak{R}}{k_B T}) \right\}^{-1} \tag{4.57}$$

We would like once more to underline that for finite systems the stationary solution of the master equation is held only for $J^S = 0$. With this condition the equilibrium distribution can be found from eq. (4.53) in the form

$$P^O(n) = P^O(1) \prod_{j=2}^{n} \frac{w_{j-1}^+}{w_j^-} \tag{4.58}$$

$P^O(n)$ gives the probability to find the cluster in equilibrium with a size n. Fig. 4.7 shows $P^O(n)$ for different initial supersaturations. It demonstrates that for small initial supersaturations the probability to find the cluster as a part of the vapour (that means $n_{eq} < n_{cr}$) becomes important (compare also Fig. 4.6).

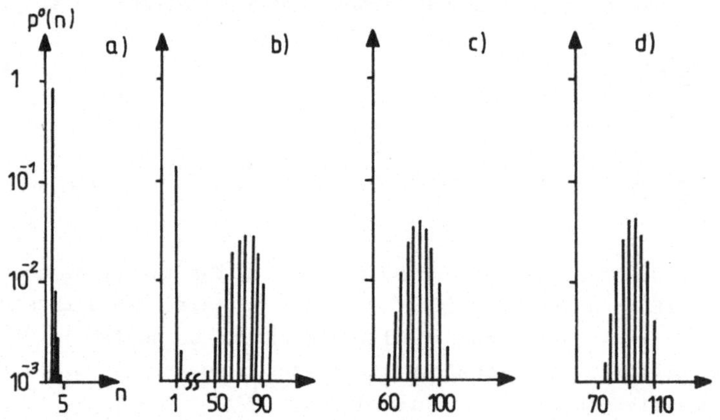

Fig. 4.7:
Probability $P^O(n)$ for the equilibrium cluster size n for different initial supersaturations
(a) $y_o = 3.86$; (b) $y_o = 4.63$; (c) $y_o = 5.15$; (d) $y_o = 5.79$
overall particle number N = 150

(From: SCHWEITZER and SCHIMANSKY-GEIER, 1987)

4.3.3. Evolution of the Cluster Distribution

We now investigate the stochastic evolution of the cluster distribution
in the finite system by means of the stochastic dynamics. The transition probabilities to change the distribution via the reaction (4.21)
are given by the eqs. (4.38) and (4.41). To simulate the whole process
of formation and growth of clusters we choose a constant overall particle number N = 150. The thermodynamic constrains are represented in
terms of the initial supersaturation y_0 = $Nk_BT/p'\cdot V$. With respect to
the constant value of N an increase of y_0 means a decrease of the temperature T or a decrease of the system volume V.
The specific properties of the vapour and the liquid phase are obtained for ethanol. For a temperature T = 290 K the constants A (eq.(4.14))
and B (eq. (4.15)) representing these properties are hold as: A =
19.08 k_BT and B = 5.32 k_BT.
The stochastic simulation starts with the restriction that in the initial state only free particles exist in the system, that means:
$P(\underline{N},t=0) = \delta_{N,N_1}$. The mean life time (eq. 4.45) of the initial state
is given in accordance with eq. (4.39):

$$\tau_{N=N_1} \sim V/N(N-1)$$

For large initial densities N/V τ becomes very short.
The first possible reaction is the formation of a dimer. For the second reaction three possibilities exist: the dimer grows up to a tri-
mer, or the dimer splits, or a second dimer will be created. One of
these reaction is randomly choosed as explained in Chapter 4.3.1. With
the obtained new particle configuration the evolution will be conti-
nued.
The results of the stochastic cluster evolution are presented in Fig.
4.8. It gives five shanpshots of the particle configuration for different moments.
The Figs. 4.9 and 4.10 shall be accompanied the discussion of the evolution of the cluster distribution. There are presented time dependent
evolution of the whole number of clusters ($n \geqslant 2$) and of the whole number of bound in clusters particles.
We can distinguish between three different stages of the phase transition.
(i) First a period of predominant nucleation occurs where a big number
of undercritical clusters is formed and a distribution of free particles, dimers, trimers have been established in a very short ,time
(after about 500 reactive collisions). The number of clusters increases
rapidly and the number of bound particles too (compare Fig. 4.8a).

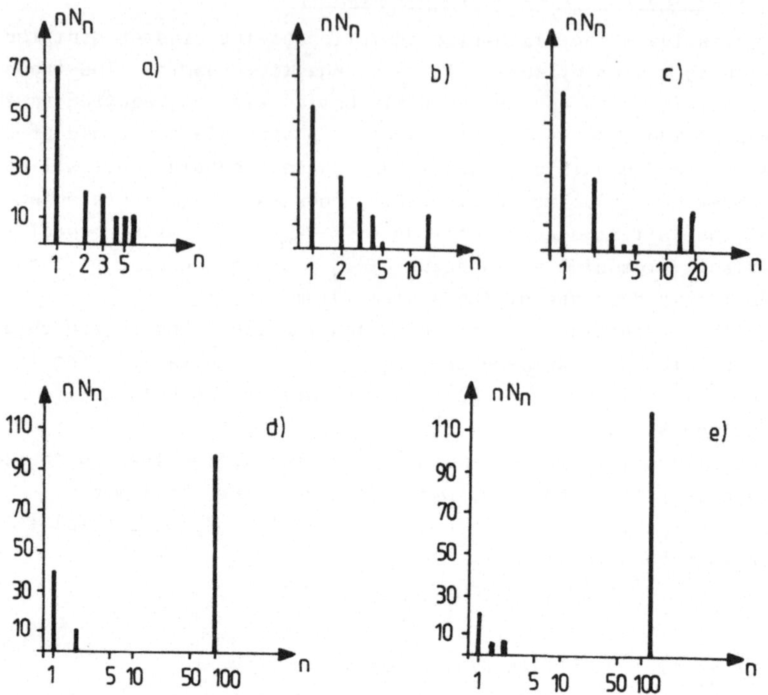

Fig. 4.8:
Stochastic evolution of the cluster distribution. There is presented the number nN_n versus cluster size n for different time moments (in time units). R gives the number of reactive collisions during the given time t, y denotes the recent supersaturation in the system. Initial supersaturation y_o = 12.5

(a) R = $5 \cdot 10^2$ t = 0.368   y = 8.35

(b) R = $1 \cdot 10^3$ t = 0.719   y = 8.15

(c) R = $2 \cdot 10^3$ t = 1.472   y = 7.50

(d) R = $5 \cdot 10^3$ t = 6.750   y = 3.17

(e) R = $1 \cdot 10^4$ t =27.250   y = 2.17

(From: SCHWEITZER, 1986a)

(ii) After a certain time lag some of these small clusters begin to
grow until a supercritical size caused by larger fluctuations. In
this period a favoured cluster growth is obtained. The number of
bound particles increases again, but the number of clusters already
decreases. That means the nucleation period has been finished (compare
Fig. 4.8 b,c).

(iii) In the last but the longest period the number of clusters and
the number of bound particles are nearly constant, but fluctuate because always new dimers are creating or disappear. During this period
one of the larger clusters grows to its final state. This cluster succeeds due to a competition process, that means its further growth occurs at the expence of the smaller clusters which have to disappear.
This process is also called "Ostwald ripening", characterized by a re-
store of the already bound particles from smaller to larger clusters
caused by the re-evaporation of the small clusters (compare Fig. 4.8,
d,e).
The final state (Fig. 4.8 e) is given by one large cluster surrounded
by a distribution of mainly free particles and small clusters. This
state is in accordance with the thermodynamic equilibrium of the two
coexisting phases. The stable cluster size is determined by the thermodynamic constraints only. It can be calculated approximately from
the Gibbs-Thomson equation (3.27).
We note again, that the process of formation of critical clusters, (i),
can be understood only by the consideration of fluctuations in the system, while the process of growth of supercritical clusters and the
Ostwald ripening period are able to describe by deterministic equations. The Chapters 5 and 6 are devoted to a detailed investigation of
this problem.
The time dependence of the supersaturation in the system presented in
Fig. 4.11 confirms the given results. The initial supersaturation y_o
is reduced first in a very short time due to the nucleation process
and then it decreases because of the depletion of the vapour when the
overcritical clusters grow. In the last period the pressure in the system is nearly constant and changes stochastically.

From   other simulation experiments we obtain the time dependent behaviour of the free energy of the cluster distribution in comparison
with the free energy of the initial state. In Fig. 4.12 it is found an
increase of the free energy in the period of formation of the critical
clusters. The maximum of ΔF gives the already discussed nucleation
barrier which has to cross over. While the overcritical clusters grow
the free energy decreases again. In the final state it has reached a
minimum value lower than the free energy of the initial supersaturated
state.
Fig. 4.13 presents the relationship between the number of reactive
collisions and the clock of the system given by the sum over the actual
life times of the different cluster distributions. To find a suitable
time scale the parameter α has been determined in comparison with

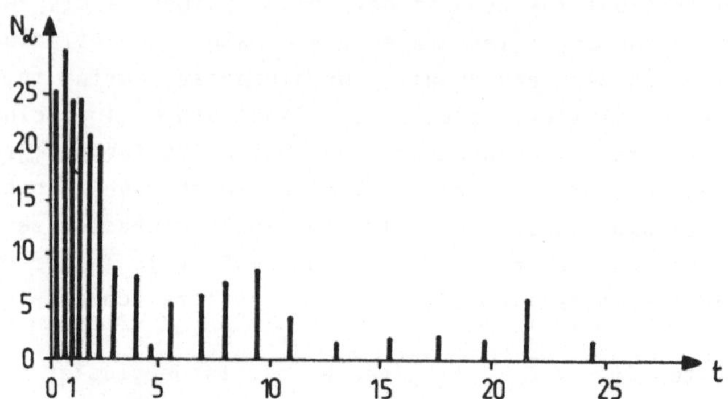

Fig. 4.9:
Stochastic evolution of the number of clusters $N_\alpha = \sum_{n=2}^{N} N_n$ versus time
(in time units). Initial supersaturation $y_0 = 12.5$
(From: SCHWEITZER, 1986a)

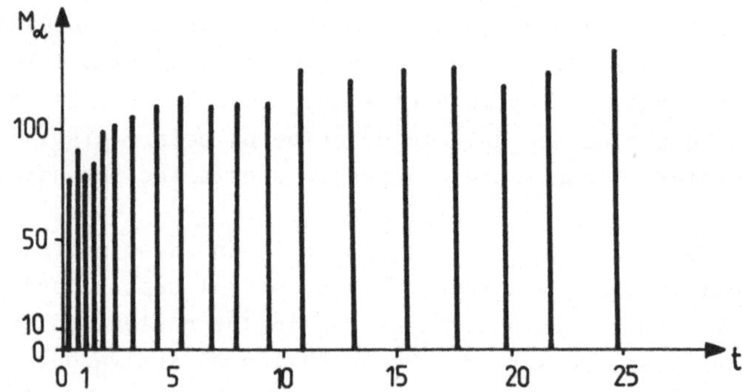

Fig. 4.10:
Stochastic evolution of the number of bound in cluster particles $M_\alpha = \sum_{n=2}^{N} n N_n$ versus time (in time units). Initial supersaturation $y_0 = 12.5$
(From: SCHWEITZER, 1986a)

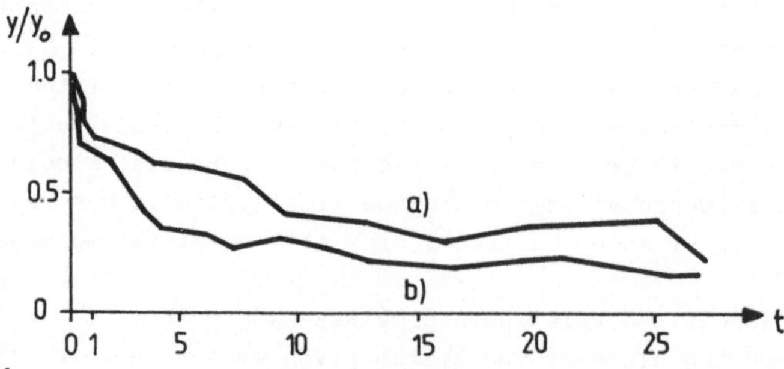

Fig. 4.11:
Stochastic evolution of the reduced supersaturation $y(t)/y_0$ versus time
(in time units). Initial supersaturation: (a) $y_0 = 12.5$; (b) $y_0 = 7.5$
(From: SCHWEITZER, 1986a)

108

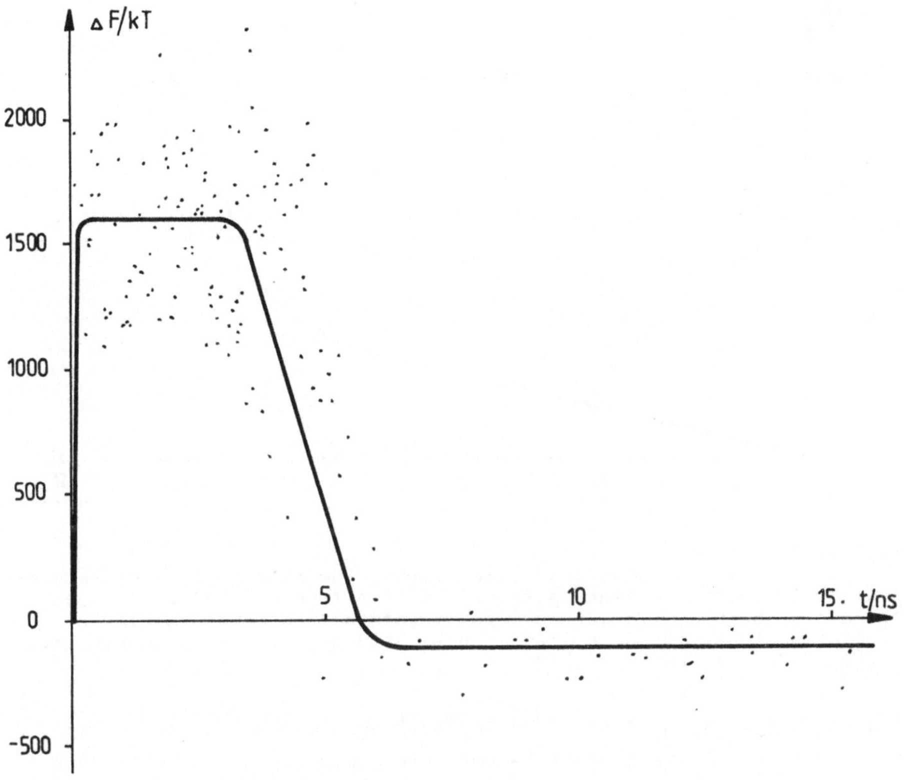

Fig. 4.12:
Change of the free energy $\Delta F(T,V,\underline{N}) = F(T,V,\underline{N}) - F_{hom}$ of the cluster distribution compared with the value of the initial state ($F_{hom}/kT -$ $- 4995$) versus time (in nanoseconds). The dots represent numerical results coming from several computer runs; the solid line is a smooth interpolation curve.
vapour: water; initial supersaturation: $y_o = 18$; temperature: $T = 293$ K; overall particle number: $N = 300$

(From: MAHNKE and BUDDE, 1988)

known deterministic equations as $\alpha = D/l_o$ (MAHNKE and BUDDE, 1987). The characteristic length l_o is defined in eq. (4.95). The increase of the time in the range of large numbers of collisions means an increasing life time of the established cluster distribution near the thermodynamically stable state.

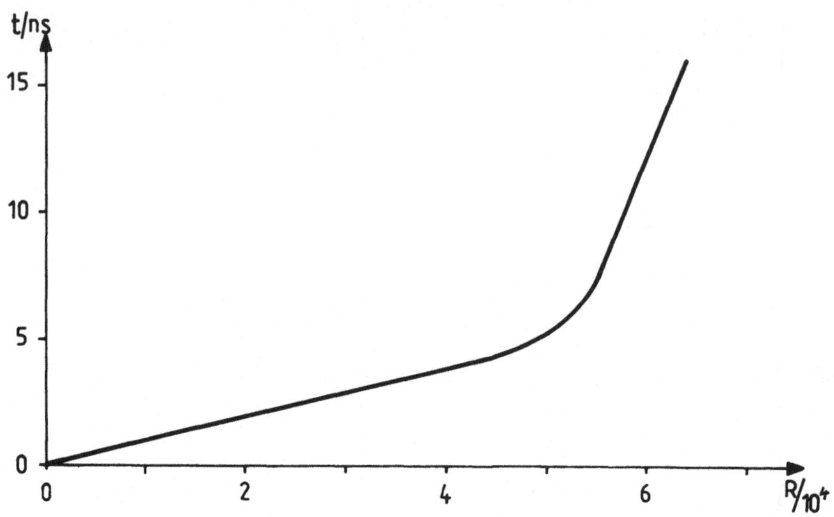

Fig. 4.13:
Scaled time of the phase transition versus the number of reactive collisions in supersaturated water vapour. The frequency of collisions
changes from $10^4 ns^{-1}$ to $10^3 ns^{-1}$ which demonstrates the symmetry
breacking effect in phase transitions. Same parameters as in Fig. 4.12.
(From: MAHNKE and BUDDE, 1988)

The stochastic simulation of the phase transition due to nucleation
and growth of clusters drafts a very clear picture of the whole process. The results are presented here only for a single run. To find
statements for the mean value's behaviour we have to average over a big
number of independent runs. But the improvement of the results does
not change the general statements for the time dependent evolution of
the phase transition.
The results of simulation agree with the outlined scenario of the phase transition given in the Chapters 5 and 6. There some special aspects
of the cluster growth will be clarified which is based on the thermodynamic results of Chapter 3.

4.4. Probability Distribution and Mean First Passage Time

To get a deeper insight into the stochastic evolution of the cluster
distribution in finite systems we complete the given investigations in
this Chapter with a discussion of the probability distribution and the
mean first passage time.

Let $p(n,t)$ denote the probability to find any cluster with a size n
at a given time t in the system. For the initial state which consists
of only free particles, $p(n,t=0)$ is a delta function in the state $n=1$.
The time dependent evolution of $p(n,t)$ is sketched in Fig. 4.14. It can
be described by means of two time scales (JANSSEN, 1974; OPPENHEIM
et al., 1977, EBELING and SCHIMANSKY-GEIER, 1979, 1980):

(i) In a shorter, quasistationary time scale the initial probability
distribution relaxes into a Poissonian distribution around the nearest
stable state. This state is given by the vapour phase being a metastable one. The metastable cluster distribution consisting of only small
clusters and free particles exists in the supersaturated state over a
certain time. That means the behaviour during the quasistationary time
scale agrees with the deterministic picture where the metastable state
can not be left.

(ii) During a larger, stationary time scale overcritical fluctuations
become important. The metastable state is left and the probability
distribution will be broaden. Near the second stable·state (drop) another maximum of the probability distribution is built up. Thus the
phase transition occurs in the stationary time scale which consider
the stochastic behaviour.

At the end of the phase transition we find instead of the metastable
unimodal probability distribution a bimodal distribution, which describes the coexistence of a stable cluster and the vapour by means of
two maxima of $p(n,t)$.

The final state is asymptotically reached. In this case the maxima of
$p(n,t \rightarrow \infty)$ agree with the thermodynamic stable states of the deterministic picture, while the minimum is related to the instable state.

The discussion of the probability distribution is illustrated by
means of computer simulations (SCHWEITZER, 1986a, 1987). Fig. 4.15 presents instead of the probability distribution the frequency to find
any cluster with size n during a single run. For a big number of independent runs the frequency distribution converges into the probability distribution. It is to be seen that for less than 0.5 time units
the metastable state is not left. Afterwards the frequency to find larger clusters becomes important. During the phase transition a broad
distribution of the frequency is obtained. When the stable cluster size
is reached the frequency is confined near the second stable state and

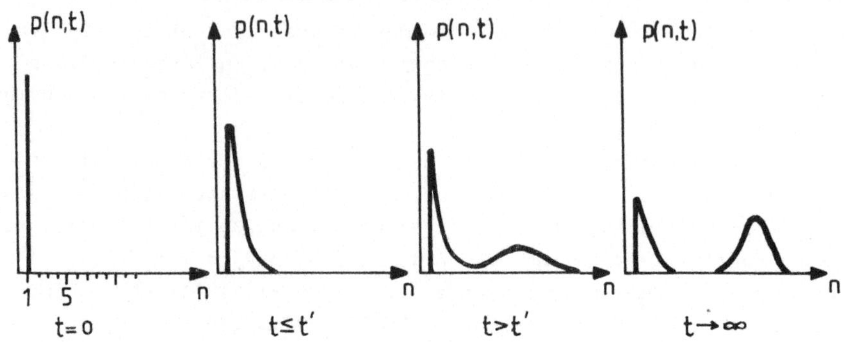

Fig. 4.14:
Sketch of the probability distribution to find any cluster of size n
for different moments. t' denotes the mean first passage time to
reach a state where overcritical clusters ($n > n_{cr}$) exist.

for large times we find a clear bimodal distribution. The fluctuation
of the stable cluster size can be found easy from this Figure.
To characterize the two time scales the concept of the mean first passage time can be used. The basic ideas were already outlined fourty
years ago (KRAMERS, 1940; MONTROLL and SHULER, 1958; OPPENHEIM et al.,
1977), but this concept still requires a recent theoretical interest
in the field of discrete Markov processes (LINDENBERG, SESHADRI, 1979;
WEST et al., 1980; MATKOVSKY et al., 1984; KNESSL et al., 1984; 1986;
HÄNGGI, 1986; WEISS, 1986). It can be applicated also to a lot of
practical problems in electronics on chemical kinetics (MATHESON et
al., 1978; FLEMING et al., 1986; HYNES, 1986).
The mean first passage time into the critical state gives a measure
for the quasistationary time scale. It approximates the life time of
the metastable cluster distribution.
With the formation of the first critical cluster the probability of
the phase transition increases. The minimum time to establish the bimodal probability distribution can be characterized by the mean first
passage time into the stable state. Therefore it is a measure for the
stationary time scale where the phase transition occurs.
Tab. 4.1 gives some values for the mean first passage times into the
critical and the stable state averaged over 20 independent computer
runs. With an increasing initial supersaturation the transition times
are shorten, but they fluctuate very strong.

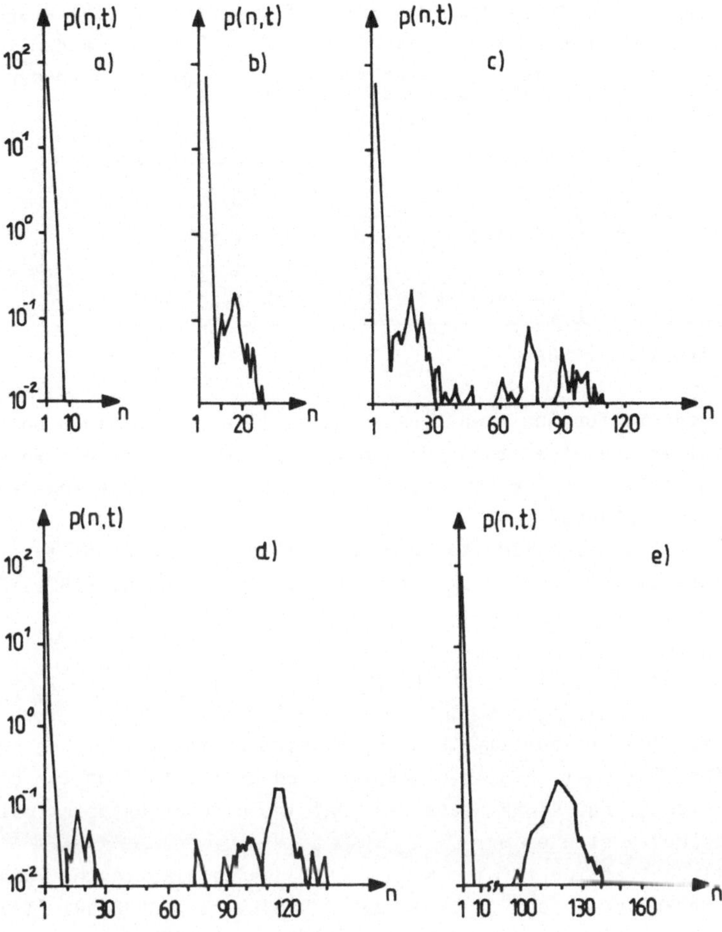

Fig. 4.15:
Frequency to find clusters with size n during a given time t (in
time units) for a single run. R denotes the number of reactive collisions during the given time:

(a) R = $5.0 \cdot 10^2$;   t =   0.368

(b) R = $2.5 \cdot 10^3$;   t =   1.931

(c) R = $5.0 \cdot 10^3$;   t =   6.752

(d) R = $1.0 \cdot 10^4$;   t = 27.250

(e) R = $3.2 \cdot 10^4$;   t =129.371

initial supersaturation y_o = 12.5
vapour: ethanol, N = 150, T = 290 K

(From: SCHWEITZER, 1986a)

Tab. 4.1:
Mean first passage time (in time units) from the initial state into
the critical state (τ_{cr}^+) and into the stable state (τ_{st}^+) averaged
over 20 runs. y_o denotes the initial supersaturation, n_{cr} and n_{st} are
the critical and the stable cluster sizes for the given thermodynamic
constraints, σ denotes the standard deviation
vapour: ethanol, N = 150, T = 290 K

y_o	n_{cr}	n_{st}	$\tau_{cr}^+ \pm \sigma_{cr}$	$\tau_{st}^+ \pm \sigma_{st}$
7.5	13	108	3.03 ± 2.64	21.24 ± 11.96
10.0	10	119	0.81 ± 0.43	11.30 ± 3.02
12.5	7	125	0.29 ± 0.18	9.23 ± 1.62

(From: SCHWEITZER, 1986a)

Analytic results for the mean first passage time are obtained only in
the case of an one-dimensional random walk process. Therefore we discuss in the following the evolution of a single cluster in the finite
system like in Chapter 4.3.2.
The mean first passage time to reach a state $n > m$ that means to find
the cluster with size $n > m$ in the system can be given by (EBELING
et al., 1977):

$$\tau = \int_o^\infty Q(t)\,dt \ , \quad Q(t) = \sum_{j=1}^m P(j,t) \tag{4.59}$$

$Q(t)$ denotes the probability that the cluster at the time t has not
reached the size $n > m$, $P(j,t)$ gives the probability to find the cluster
with the size j. Eq. (4.59) supposes that the cluster asymptotically
surely reached a size $n > m$.
An analytic expression for the mean first passage time in the case of
discrete Markov processes is given by LINDENBERG and SESHADRI (1979).
This expression is hold for the asymptotic limit, therefore it depends
on the equilibrium probability $P^o(n)$ given by eq. (4.58):

$$\tau^+(n+1 \mid m) = \sum_{j=m}^n \frac{1}{w_j^+ P^o(j)} \sum_{i=1}^j P^o(i) \ , \quad m \leq n \tag{4.60}$$

For the case of re-evaporation of the cluster from size n to a size
$k < n$ LINDENBERG and SESHADRI (1979) find the expression:

$$\tau^-(k \mid n) = \sum_{j=k+1}^n \frac{1}{w_j^- P^o(j)} \sum_{i=j}^N P^o(i) \quad ; \quad n \geq k \tag{4.61}$$

w_j^+ and w_j^- are the transition probabilities in the given directions.
For the considered case we use eq. (4.51). The results for the mean
first passage time, eqs. (4.60), (4.61) are concluded in Tab. 4.2.

Tab. 4.2:
Mean first passage times (unnormalized) from the initial cluster size
($n=1$) into the critical size (n_{cr}) and into the stable cluster size
(n_{st}) (eq. 4.60)
$\tau^-(1|n_{st})$ (eq. (4.61)) denotes the mean first passage time for a re-
evaporation of the cluster assumed to be in the stable state.
y_0 gives the initial supersaturation. vapour: ethanol, $N = 150$, $T = 290$ K

y_0	n_{cr}	n_{st}	$\tau^+(n_{cr}|1)$	$\tau^+(n_{st}|1)$	$\tau^-(1|n_{st})$
3.86	34	44	$7.36 \cdot 10^5$	$1.12 \cdot 10^6$	$3.23 \cdot 10^1$
4.21	22	63	$5.17 \cdot 10^4$	$1.19 \cdot 10^5$	$3.62 \cdot 10^2$
4.63	19	75	$7.23 \cdot 10^3$	$1.66 \cdot 10^4$	$2.51 \cdot 10^7$
5.15	13	85	$1.60 \cdot 10^3$	$3.33 \cdot 10^3$	$2.51 \cdot 10^7$
5.79	11	93	$3.72 \cdot 10^2$	$8.30 \cdot 10^2$	$2.07 \cdot 10^{11}$

(From: SCHWEITZER, 1986a)

The mean first passage times can be used to define a measure for the
stability of the cluster states (PROCACIA and ROSS, 1977). Large
transition times into other states indicate a stable initial state.
For large initial supersaturations we observe a short passage time
into the stable cluster state, but a large time to leave this state
again (see also SCHWEITZER et al., 1987). The probability for the
cluster to re-evaporate is very small in this case (compare also Fig.
4.6 in Chaptor 4.3.2) With an increasing initial supersaturation the
mean first passage times from $n=1$ into the critical and the stable
state decrease obviously. An important contribution (about the half
of the value) of the mean first passage time into the stable state is
given by the time to reach the critical state and to cross over the
nucleation barrier. This allows us to approximate the mean first passage time by the contribution to overwhelming the critical state only.
A suitable approximation for this case was given by WEST et al.
(1980):

$$\tau^+(m+1|n) \sim \frac{1}{w_{n_{cr}}^+ \, P^0(n_{cr})} \sum_{j=1}^{n_{cr}} P^0(j) \ , \quad n < n_{cr} < m \tag{4.62}$$

$$\tau^-(k|n) \sim \frac{1}{w_{n_{cr}}^- \, P^0(n_{cr})} \sum_{j=n_{cr}}^{N} P^0(j), \qquad n > n_{cr} > k \tag{4.63}$$

Another form to estimate the stability of the different cluster states should be finally discussed. If a stable cluster size $(n > n_{cr})$ exists for the given thermodynamic constraints (see also Chapter 3) we obtain a bimodal equilibrium probability distribution $P^0(n)$. We now reduce the information of $P^0(n)$ by introduction of a two-spin-state description (GARDINER, 1984). Because the critical cluster size decides whether the cluster will be a part of the vapour $(n < n_{cr})$ or not, we define the two states as follows:

$$s = 0 \text{ for } 1 < n < n_{cr}; \quad s = 1 \text{ for } n_{cr} < n < N$$

The equilibrium probability to find the cluster in the states $s = 0$ or $s = 1$ is then given by

$$Q_{s=0} = \sum_{j=1}^{n_{cr}} P^0(j) \; ; \quad Q_{s=1} = \sum_{j=n_{cr}+1}^{N} P^0(j) \tag{4.64}$$

The mean life times for the cluster in the states $s = 0, 1$ can be obtained in the form (SCHIMANSKY-GEIER and EBELING, 1983):

$$\gamma_{s=0} = \sum_{j=1}^{n_{st}-1} \frac{1}{w_j^+ P^0(j)} \; Q_{s=0}$$

$$\gamma_{s=1} = \sum_{j=2}^{n_{st}} \frac{1}{w_j^- P^0(j)} \; Q_{s=1} \tag{4.65}$$

γ_0 and γ_1 give a measure for the stability of the vapour phase and the cluster phase. Like in the case of the mean first passage times the main contributions in eq. (4.65) are hold for the critical cluster state. If we only consider these contributions, eq. (4.65) agree with the approximations (4.62), (4.63). Thus we find a close relation between the mean first passage time from the metastable to the stable cluster state and the mean life time of these states indicating the stability. A numerical calculation of eq. (4.65) shows indeed only a negligible deviation from the exact results obtained from the eqs. (4.60), (4.61) (SCHWEITZER, 1986a).

4.5. Mean Values for the Number of Clusters - Fokker-Planck-Equation

This Chapter deals with a discussion of the mean values of the number of clusters N_n. If we know the kinetic equation for the mean values which will be given here by a Fokker-Planck equation then it is easy to derive the deterministic equations in the limit of vanishing fluctuations. Therefore the discussion of the mean values leads up from the stochastic description to the deterministic considerations given in the following Chapters.

The mean number of clusters of size n is received from the first moment of the probability $P(\underline{N},t)$:

$$\langle N_n(t) \rangle = \sum_{\{N_i\}} N_n P(N_1 \ldots N_n \ldots N_N, t) \tag{4.66}$$

$\{N_i\}$ means every possible cluster distribution which fulfilles the restrictive condition N = const.

The time dependent change of the mean values $\langle N_n(t) \rangle$ is given by:

$$\frac{d}{dt} \langle N_n(t) \rangle = \sum_{\{N_i\}} N_n \frac{\partial P(N,t)}{\partial t} \tag{4.67}$$

By means of the master equation (4.23) this equation can be written in the form (EBELING and FEISTEL, 1982):

$$\frac{d}{dt} \langle N_n(t) \rangle = \sum \Delta_j N_n \langle w_j(\underline{N}'|\underline{N}) \rangle \tag{4.68}$$

$\Delta_j N_n$ gives the value of the change of N_n for every possible reaction j (eq. (4.21)) where the N_n participate. $\langle w_j(\underline{N}'|\underline{N}) \rangle$ denotes the mean value of the related transition probability for the reaction j.

Using the transition probabilities (eqs. (4.38), (4.41)) we obtain from eq. (4.68) the following system of equations for the time evolution of the mean values (SCHIMANSKY-GEIER et al., 1986):

$$\frac{d}{dt} \langle N_n \rangle = \langle -w_n^-(N_n) - w_n^+(N_1 N_n) + w_{n+1}^-(N_{n+1}) + w_{n-1}^+(N_1 N_{n-1}) \rangle$$

$$n = 2, \ldots, N \tag{4.69}$$

$$\frac{d}{dt} \langle N_1 \rangle = \langle -2w_1^+(N_1) + 2w_2^-(N_2) - \sum_{j=2}^{N} \{ w_j^+(N_1 N_j) - w_{j+1}^-(N_{j+1}) \} \rangle$$

$$= -\langle w_1^+(N_1) + w_2^-(N_2) + \sum_{j=1}^{N} \{ w_j^+(N_1 N_j) - w_{j+1}^-(N_{j+1}) \} \rangle \tag{4.70}$$

Treating the terms of the r.h.s. of eq. (4.69) as a discrete derivation with respect to n, eq. (4.69) can also be written in the form:

$$\frac{d}{dt} \langle N_n \rangle = \langle \frac{\partial}{\partial n} \{ w_{n+1}^- (N_{n+1}) - w_n^+ (N_1 N_n) \} \rangle \tag{4.71}$$

Further it holds from eq. (4.69) for the sum of all numbers of clusters $(n \geqslant 2)$:

$$\frac{d}{dt} \sum_{n=2}^{N} \langle N_n \rangle = \langle w_1^+ (N_1) - w_2^- (N_2) \rangle \tag{4.72}$$

This implies in particular that the mean number of all clusters $(n \geqslant 2)$ can be changed only by the creation or evaporation of dimers. If we exclude this process the whole number of clusters is constant.

For the whole number of particles it follows obviously in the considered case:

$$\frac{d}{dt} \left\{ \langle N_1 \rangle + \sum_{n=2}^{N} \langle n N_n \rangle \right\} = 0 \tag{4.73}$$

If we introduce a mean supersaturation $\langle y(t) \rangle$ in analogy to the initial supersaturation y_0 in the form:

$$\langle y(t) \rangle = \frac{k_B T}{p' V} \sum_{n=1}^{N} \langle N_n(t) \rangle \tag{4.74}$$

then the time dependent change of the supersaturation is received from:

$$\frac{d}{dt} \langle y(t) \rangle = - \frac{k_B T}{p' V} \sum_{n=1}^{N} \langle w_n^+ (N_1 N_n) - w_{n+1}^- (N_{n+1}) \rangle \tag{4.75}$$

The sum in eq.(4.75) should be positive, that indicates the mean supersaturation is decreasing with time.

To investigate the time dependence of the mean numbers of clusters we now derive a Fokker-Planck equation for the mean values. The variables N_n are treated as a continuous function of n, further it is assumed that stochastic transitions are possible only between states of a close vicinity.

Starting with eq. (4.61) we make use of a Taylor expansion for the transition probabilities:

$$w_{n+1}^- (N_{n+1}) = w_n^- (N_n) + \frac{\partial}{\partial n} w_n^- \{n+1-n\} + \frac{1}{2} \frac{\partial^2}{\partial n^2} w_n^- \{n+1-n\}^2 + \ldots$$

$$w_{n-1}^+ (N_1 N_{n-1}) = w_n^+ (N_1 N_n) + \frac{\partial}{\partial n} w_n^+ \{n+1-n\} + \frac{1}{2} \frac{\partial^2}{\partial n^2} w_n^+ \{n-1-n\}^2 + \ldots$$

$$\tag{4.76}$$

Neglecting terms of higher than the second derivative we get with
eq. (4.76) from eq. (4.69) the Fokker-Planck equation:

$$\frac{d}{dt}\langle N_n\rangle = -\left\langle\frac{\partial}{\partial n}\left\{w_n^+(N_1 N_n) - w_n^-(N_n)\right\}\right\rangle$$

$$+ \frac{1}{2}\left\langle\frac{\partial^2}{\partial n^2}\left\{w_n^+(N_1 N_n) + w_n^-(N_n)\right\}\right\rangle \tag{4.77}$$

Using the transition probabilities (eqs. (4.38), (4.41)) with the
approximation $\langle N_1 N_n\rangle \approx \langle N_1\rangle\langle N_n\rangle$ we receive the following Fokker-Planck
equation for the mean cluster distribution (SCHIMANSKY-GEIER et al.,
1986):

$$\frac{d}{dt}\langle N_n\rangle = -\frac{\partial}{\partial n}\left\{\alpha\langle n\rangle^{2/3}\langle N_n\rangle\left[\frac{\langle N_1\rangle}{V} - \frac{p'}{k_B T}\exp\left(\frac{2}{3}\frac{B}{k_B T}\langle n\rangle^{-1/3}\right)\right]\right\}$$

$$+ \frac{1}{2}\frac{\partial^2}{\partial n^2}\left\{\alpha\langle n\rangle^{2/3}\langle N_n\rangle\left[\frac{\langle N_1\rangle}{V} + \frac{p'}{k_B T}\exp\left(\frac{2}{3}\frac{B}{k_B T}\langle n\rangle^{-1/3}\right)\right]\right\} \tag{4.78}$$

The time dependent change of the mean numbers of clusters of size
n ≥ 2 is now described in terms of a differential equation. The change
of the mean number of free particles must be considered with respect to
the conservation of the overall particle number, expressed by eq.
(4.73).
In a compact form the Fokker-Planck equation can be written as follows:

$$\frac{d}{dt}\langle N_n(t)\rangle = -\frac{\partial}{\partial n}\left[v_n\langle N_n(t)\rangle - \frac{\partial}{\partial n}\left(a_n\langle N_n(t)\rangle\right)\right] \tag{4.79}$$

with the quantities (SCHWEITZER, 1986b):

$$v_n = \alpha\langle n\rangle^{2/3}\left[\frac{\langle N_1\rangle}{V} - \frac{p'}{k_B T}\exp\left\{\frac{2}{3}\frac{B}{k_B T}\langle n\rangle^{-1/3}\right\}\right] \tag{4.80}$$

and

$$a_n = \frac{1}{2}\alpha\langle n\rangle^{2/3}\left[\frac{\langle N_1\rangle}{V} + \frac{p'}{k_B T}\exp\left\{\frac{2}{3}\frac{B}{k_B T}\langle n\rangle^{-1/3}\right\}\right] \tag{4.81}$$

We devide two characteristic terms in the Fokker-Planck equation
(4.79): the drift term $v_n\langle N_n\rangle$ and the diffusion term $a_n\langle N_n\rangle$. The drift
term describes the deterministic behaviour of the mean cluster distri-
bution; while the diffusion   term expresses the non-deterministic behaviour with respect to the fluctuations in the system.
v_n (eq. (4.80)) is interpreted to be the mean velocity of the deterministic cluster growth and shrinkage and will be discussed later. a_n
(eq. (4.81)) is a diffusion   parameter indicating the measure of the
fluctuations. It can be approximated by the diffusion constant D.

The stationary solution of the Fokker-Planck equation is obtained from the condition

$$\frac{d}{dt} \langle N_n \rangle = 0$$

We find the relation

$$\left[v_n + \frac{\partial}{\partial n} a_n \right] \langle N_n \rangle^0 = a_n \frac{\partial}{\partial n} \langle N_n \rangle^0 \tag{4.82}$$

with $\langle N_n \rangle^0$ being the stationary cluster distribution. For finite systems it agrees with the equilibrium cluster distribution discussed already in eq. (4.34).

It follows from eq. (4.82) that the kinetic coefficients v_n and a_n cannot be stipulated independently from each other. In agreement with eq. (4.34) we can write the equilibrium cluster distribution in the form:

$$\langle N_n \rangle^0 \sim \exp\left\{ - \frac{\Delta F_n}{k_B T} \right\} \tag{4.83}$$

with

$$\Delta F_n = - \langle n \rangle \, \ln \frac{\langle N_1 \rangle k_B T}{p' V} + \frac{B}{k_B T} \langle n \rangle^{2/3}$$

being the formation energy of the cluster of size n. ΔF_n has got a maximum value for the critical cluster size n_{cr}. From $\partial \Delta F_n / \partial n = 0$ we obtain:

$$\langle n_{cr} \rangle^{1/3} = \frac{2}{3} \frac{B}{k_B T} \left(\ln \frac{\langle N_1 \rangle k_B T}{p' V} \right)^{-1} \tag{4.84}$$

With eq. (4.83) we find in the stationary case the following condition for the kinetic coefficients (SCHWEITZER, 1986b):

$$v_n + \frac{\partial}{\partial n} a_n = - \frac{a_n}{k_B T} \frac{\partial \Delta F_n}{\partial n} \tag{4.85}$$

Use of eq. (4.82) leads to the close form of the Fokker-Planck equation (4.79):

$$\frac{d}{dt} \langle N_n(t) \rangle = \frac{\partial}{\partial n} \left\{ a_n \langle N_n \rangle^0 \frac{\partial}{\partial n} \frac{\langle N_n(t) \rangle}{\langle N_n \rangle^0} \right\} \tag{4.86}$$

Finally we want to discuss the deterministic growth and shrinkage of
the mean numbers of clusters. Neglecting the diffusion term $a_n \langle N_n \rangle$
of eq. (4.79) considering the fluctuations we derive the following
Liouville equation:

$$\frac{d}{dt} \langle N_n(t) \rangle = - \frac{\partial}{\partial n} \left\{ \alpha \langle n \rangle^{2/3} \langle N_n \rangle \left[\frac{\langle N_1 \rangle}{V} - \frac{p'}{k_B T} \exp\left(\frac{2}{3} \frac{B}{k_B T} \langle n \rangle^{-1/3}\right) \right] \right\}$$

$$n \geqslant 2 \qquad\qquad (4.87)$$

Eq. (4.87) does not take into account the formation or disappearing
of dimers. Therefore it can be written in form of a continuous equation:

$$\frac{d}{dt} \langle N_n(t) \rangle + \mathrm{div} \left[\langle N_n(t) \rangle \cdot \dot{n} \right] = 0 \qquad\qquad (4.88)$$

In general the formation or disappearing of dimers has to be considered by a source term on the r.h.s. of eq. (4.88).
Comparing the eqs. (4.87) and (4.89) we find the deterministic equation for the time dependent change of the mean cluster size n :

$$\frac{d}{dt} \langle n \rangle = \alpha \langle n \rangle^{2/3} \left[\frac{\langle N_1 \rangle}{V} - \frac{p'}{k_B T} \exp\left\{ \frac{2}{3} \frac{B}{k_B T} \langle n \rangle^{-1/3} \right\} \right] \qquad (4.89)$$

To explain this equation we first use the power expansion:

$$\exp\left(\frac{2}{3} \frac{B}{k_B T} \langle n \rangle^{-1/3}\right) = 1 + \frac{2}{3} \frac{B}{k_B T} \langle n \rangle^{-1/3} + \ldots$$

$$\ln\left(\frac{\langle N_1 \rangle k_B T}{p' V}\right) = \frac{\langle N_1 \rangle k_B T}{p' V} - 1 + \ldots$$

Inserting this expansions into eq. (4.89) it yields:

$$\frac{d}{dt} \langle n \rangle = \alpha \langle n \rangle^{2/3} \frac{p'}{k_B T} \left[\ln \frac{\langle N_1 \rangle k_B T}{p' V} - \frac{2}{3} \frac{B}{k_B T} \langle n \rangle^{-1/3} \right] \qquad (4.90)$$

The first term in brackets can be expressed by the mean critical cluster size $\langle n_{cr} \rangle$ (eq. (4.84)). Thus we receive the deterministic kinetics in the form:

$$\frac{d}{dt} \langle n \rangle = \alpha \frac{2}{3} \frac{B}{k_B T} \frac{p'}{k_B T} \langle n \rangle^{2/3} \left[\frac{1}{\langle n_{cr} \rangle^{1/3}} - \frac{1}{\langle n \rangle^{1/3}} \right] \qquad (4.91)$$

Eq. (4.91) gives the mean velocity of the growth and shrinkage of a
cluster of size n. It agrees with eq. (4.80). The mean velocity of the
cluster growth can be alternated. It holds:

$$v_n \gtreqless 0 \; , \; \text{if} \; n \gtreqless n_{cr} \tag{4.92}$$

That means the critical cluster size acts as a selection value. Only clusters with an overcritical size are able to grow, undercritical clusters have to disappear again.

The mean critical cluster size depends on time because the number of free particles $\langle N_1 \rangle$ changes with time. It reduces with the formation and growth of clusters, that's why the critical cluster size possesses an information on the recent stage of the phase transition.

In the assumed initial state where only free particles exist in the system, the critical cluster size has got its smallest value. We find from eq. (4.84)

$$\langle n_{cr} \rangle^{1/3}(t=0) = \frac{2}{3} \frac{B}{k_B T} \, (\ln \, y_o)^{-1} \tag{4.93}$$

which agrees with the critical cluster size for an infinite system (SCHWEITZER et al., 1984b). y_o gives the initial supersaturation again.

The value n_{cr} increases during the phase transition because the number of free particles decreases (compare Fig. 4.16). When the system reaches its stable state the critical cluster size converges into the stable cluster size and only the large stable cluster succeeds. The coexistence between the stable cluster and the vapour phase is described the Gibbs-Thomson equation, which has been already discussed in Chapter 3.

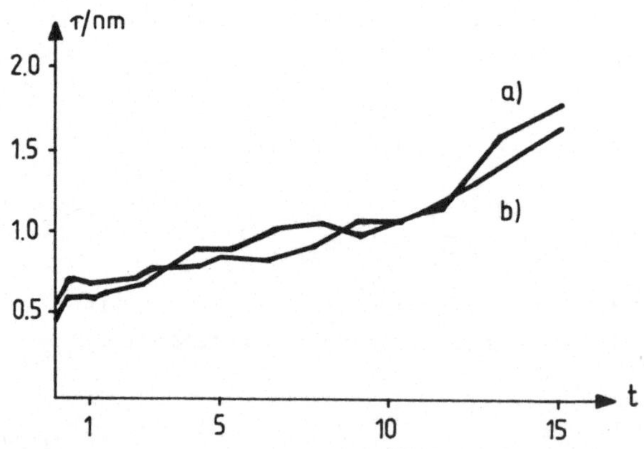

Fig. 4.16:
Stochastic evolution of the critical cluster size (in nm) versus time (in time units). Initial supersaturation: (a) y_o = 12.5; (b) y_o = 7.5; vapour: ethanol, N = 150, T = 290 K
(From: SCHWEITZER, 1986a)

For a comparison with the results given in the following Chapters wc
note finally that the deterministic equation (4.91) for the cluster
growth gets the form (SCHWEITZER, 1986b):

$$\frac{d}{dt}\, r_n = \frac{\alpha}{3}\, \left(\frac{4\pi}{3}\, c_\alpha\right)^{-1/3} \frac{p'}{k_B T}\, l_0 \left[\frac{1}{r_{cr}} - \frac{1}{r_n}\right] \tag{4.94}$$

if we express the cluster size n by the cluster radius r_n and the
constant B (eq. (4.15)) by the capillary length l_0. It yields the relation:

$$\frac{2}{3}\, \frac{B}{k_B T}\, \langle n \rangle^{-1/3} = \frac{2\sigma}{c_\alpha k_B T}\, \frac{1}{r_n} = \frac{l_0}{r_n} \tag{4.95}$$

5. Kinetics of Growth of a New Phase - A Deterministic Description

5.1. General Scenario of First-Order Phase Transition in Finite Systems

A theoretical description of nucleation not taking into account the simultaneous growth of the cluster is only possible if the depletion of the medium can be neglected as presumed in the classical nucleation theory. In this case we obtain a simultanous nucleation and independent growth of the already formed supercritical clusters.
For a finite system the formation and growth of the clusters result in a depletion of the surrounding medium. Thus we come to another scenario of the first-order phase transition in finite systems. This scenario can be deduced from the results of the stochastic description of the phase transition and the related computer simulations, given in Chapter 4, as well as from the thermodynamic investigations, given in Chapter 3. By means of this results we obtain the following general scenario:
First a process of nucleation occurs in the initial homogeneous supersaturated state requiring a very short time. In a second stage the growth of the already formed supercritical clusters predominates and the nucleation rate decreases. This stage is succeeded by a third stage of competitive growth of the clusters, the so-called Ostwald ripening period. It is characterized by a decrease of the number of clusters and an increase of their mean radius during a large interval of time. Because of the depletion of the medium the three stages are not independend from each other. In particular the nucleation rate depends on the growth of the already formed clusters.

The outlined scenario of first-order phase transitions is valid, if the initial supersaturation is not too high. In this case the first stage of the transition can be described by nucleation - the formation of fluctuations in small regions of space with large differences in the density compared with the initial staté. If the initial supersaturation is increased, nucleation is replaced continuously by spinodal decompositions (HILLERT, 1961; CAHN, HILLIARD, 1958, 1959; CAHN, 1961, 1962, 1963, 1965, 1966, 1968; COOK, 1970; LANGER, BARON, MILLER, 1975). Here the initial stages of the transition are characterized by the formation of long wave length fluctuations with relatively small differences in the density compared with the initial state, an up-hill diffusion, leading to an increase of the density differences and the formation of two distinct phases.

In general, the evolving phase is characterized by a large surface
area and, again, the two first stages of the transition are followed
by a coarsening process. If, in addition, this process proceeds mainly
via the addition of monomers, the theory of Ostwald ripening outlined
in Chapter 6 is applicable for a theoretical description.
Here already another restriction of the applicability of the outlined
scenario is mentioned, the growth of the new phase must proceed mainly
via the addition of monomers. Other mechanisms of growth like coagulation, sedimentation, hydrodynamic effects may lead to modifications of
this scenario (see, e.g., KRISHNAMURTHY, GOLDBURG, 1980; SIGGIA, 1979;
WONG, KNOBLER, 1981).

For a deeper insight in the first-order phase transition in finite systems in this chapter we complete the investigations of Chapter 4 with
some analytical results obtained from a determination description of
the process. Therefore we refer to the thermodynamic results as summarized in Fig. 3.18.

The dotted curve in Fig. 3.18 marks the stage of dominating nucleation,
and increase of the number of critical clusters accompanied in finite
systems by an increase of the critical cluster size. s_c is the maximum
number of critical drops which can be formed as a result of nucleation
(point P in Fig. 3.18). However, this maximum number is not formed,
because the process of overwhelming nucleation will go over earlier
into a second stage of a predominating and practically independent
growth of the clusters, their number being nearly constant. This stage
is marked by a dashed-dotted line in Fig. 3.18 starting at point P'.
With an increasing initial supersaturation the number of clusters increases, therefore the point P' in Fig. 3.18 moves with an increasing
initial supersaturation into the direction of P.
The stage of a relatively independent growth of the clusters, connected with a significant decrease of the supersaturation is followed by
a third stage of competitive cluster growth, leading to a decrease of
the number of clusters and an increase of their mean radius. This
stage indicated in Fig. 3.18 by a dashed curve occurs along the "valley" of the characteristic thermodynamic potential and is related to
the process of Ostwald ripening.

We now characterize these three stages from a deterministic point of
view. First a quasi-steady state approximation of the nucleation process is given for the finite systems which is closely related to the
classical nucleation theory. Then a growth equation for the super-

critical clusters will be derived applying the thermodynamics of non-reversible processes. Combining this growth equation with the quasi-steady state approximation of the nucleation period it is shown that nucleation and growth of clusters are no longer independent from each other for the phase transition in finite systems.

The process of Ostwald ripening will be investigated in detail in Chapter 6.

5.2. Nucleation in Finite Systems - The Quasi-Steady-State Approximation

As already mentioned in Chapter 1, the majority of nucleation theories lead to expressions of the form

$$I = I_0 \exp(-W/k_B T) \qquad (5.1)$$

for the steady-state nucleation rate. It gives the number of critical clusters formed in a unit of time and volume.

As it was shown in Chapters 2.1. and 3.4. the work of formation of critical clusters in finite systems can be well approximated by Gibbs' expression: $W = 1/3\,\sigma A_c$ (compare eq. (2.1)). But in the calculation of the parameters of the critical clusters depletion effects have to be taken into account. Therefore we obtain:

$$W = \frac{1}{3}\sigma(4\pi r_c^2(\varrho_\beta)) \qquad (5.2)$$

For finite systems r_c depends on the state of the surrounding the clusters medium (for one-component systems on ϱ_β, the density of the medium) and since the number of clusters increases with time, also on time.

According to the generally accepted definition nucleation is the process of formation of critical or supercritical clusters. Consequently, the path marked by a dotted curve in Fig. 3.18 represents nucleation in finite systems, the formation of critical clusters connected with the depletion of the medium and an increase of the critical cluster size.

The critical cluster radius (volume of the cluster) depends on the total mass $n_{\alpha t}$ (or the total volume $V_{\alpha t}$) concentrated in the already formed clusters. In agreement with eqs. (3.68), (3.69) we can write

$$r_c = \frac{2\sigma}{\rho_\alpha(\mu_\beta(\rho_\beta)-\mu'(\rho'))-(p_\beta(\rho_\beta)-p'(\rho'))} \tag{5.3}$$

$$\rho_\beta = \rho \, \frac{1 - n_\alpha t/n}{1 - V_\alpha t/V}$$

Therefore, in a first approximation the nucleation rate in finite systems can be expressed by eq. (5.1), where W has to be determined by eqs. (5.2) and (5.3).

This approximation we denote as quasi-steady-state approximation, since the expression for the steady-state nucleation rate (5.1) is used, but on the other hand, r_c depends on ρ_β and, therefore, on time.

Though the critical radius depends on time it can be obtained from the thermodynamic analysis, that the main thermodynamic premise of classical nucleation theory and its generalizations - the existence of a critical cluster size - remains valid also for finite systems. Therefore, the classical nucleation theory and its generalizations are applicable, may be in modified form, for a description of nucleation in finite systems (compare in contrast VOGELSBERGER, 1982, 1983).

Assuming in addition to the quasi-steady state approximation, that
- all supercritical clusters, once formed, do not disappear, again
- all clusters have nearly the same (time-dependent) radius r_c (it means, that growth processes of the already formed supercritical clusters are neglected in the nucleation period)
- the gas can be considered as a perfect one

the nucleation rate was calculated based on eqs. (5.1)-(5.3). The results are presented in Fig. 5.1.

Fig. 5.1 shows that only for relatively small supersaturations the nucleation rate in the finite system is practically constant over longer periods of time. For larger supersaturations the increase of the critical radius caused by the depletion of the vapour leads to a considerable decrease of the nucleation rate. This is due to the exponential dependence of the nucleation rate I on r_c, resulting in large variations of I for small variations of r_c.

Finally we would like to mention that the same method is applicable also to multicomponent systems. In addition time-lag effects can be taken into account (KELTON et al., 1983).

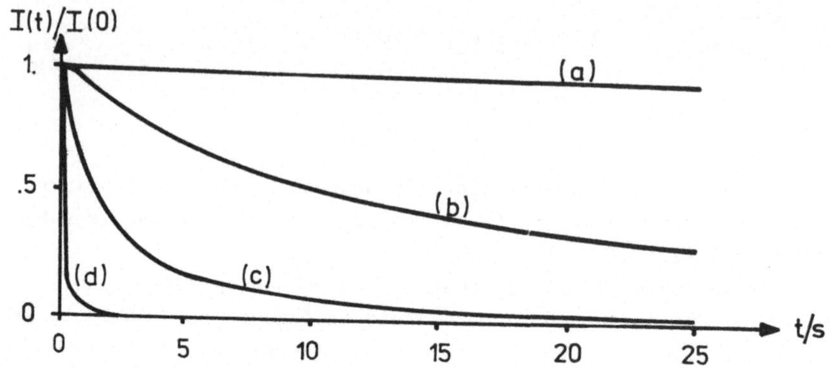

Fig. 5.1:
Nucleation rate as a function of time for different values of the initial supersaturation y_0.
a) $y_0 = 6$, b) $y_0 = 7$, c) $y_0 = 8$, d) $y_0 = 10$
The calculation is based on the quasi-steady-state approximation neglecting in addition a possible simultaneous growth of the already formed supercritical clusters.
vapor: water, $T = 293.15$ K, $I_0 = 10^{31}$ m^{-3}s^{-1}
(From: SCHMELZER, 1985a)

5.3. Deterministic Growth Equations

5.3.1. Diffusion Equation Approach

Once a supercritical cluster is formed its further development proceeds in agreement with the macroscopic laws of thermodynamics. Consequently, the thermodynamics of non-reversible processes is applicable to the description of the growth or shrinkage processes of the cluster.

The linear thermodynamics of non-reversible processes postulates a linear relationship between the thermodynamic driving forces X_i and the thermodynamic fluxes J_i of the form

$$J_i = \sum_k L_{ik} X_k \qquad (5.4)$$

In particular, if the temperature T and the pressure p are held constant throughout the system, the density of fluxes of particles of the i-th component is given by

$$\vec{J}_i = - \frac{D_i c_i}{kT} \text{ grad } \mu_i \qquad (5.5)$$

128

D_i is the partial diffusion coefficient, c_i the volume concentration
of particles, μ_i the chemical potential of the i-th component dissolved
in the medium. The eq.(5.5) can be applied to derive growth equations
for a newly formed phase.

We limit our considerations to the case, that the velocity of growth
of the cluster is determined mainly by the diffusion of one of the components. Assuming in addition, that the chemical potential of this component can be expressed by (ideal mixture)

$$\mu(p,T,x) = \mu'(p,T) + kT \ln x \tag{5.6}$$

eq. (5.5) leads to the so-called 1. Fick's law.

$$\vec{j} = - D \text{ grad } c \tag{5.7}$$

Taking into account the continuity equation, the mathematical expression for the conservation of particles,

$$\frac{\partial c}{\partial t} + \text{div } \vec{j} = 0 \tag{5.8}$$

Fick's second law results in the form

$$\frac{\partial c}{\partial t} = \text{div } (D \text{ grad } c) \tag{5.9}$$

In particular, for the spherical symmetric case this equation is reduced to

$$\frac{\partial c}{\partial t} = \frac{1}{r^2} \frac{\partial}{\partial r} \left(D r^2 \frac{\partial c}{\partial r} \right) \tag{5.10}$$

If at a distance R from a chosen centre of symmetry the concentration
of the diffusing particles is held constant ($c = c_R =$ const) and also
for $r \rightarrow \infty$ the concentration is fixed ($c = c_\infty =$ const) after some
time a stationary concentration profile is established in the system
($\partial c/\partial t = 0$). From eq. (5.10) it follows then, that this time-independent concentration profile has the following form

$$c(r) = - \frac{c_\infty - c_R}{r} R + c_\infty \tag{5.11}$$

The density of fluxes through the surface with a radius R is given
then by

$$\vec{j}_R = - D \left. \frac{\partial c}{\partial r} \right|_{r=R} \vec{e}_r = - D \frac{c_\infty - c_R}{R} \vec{e}_r \tag{5.12}$$

$\vec{e}_r$ is a unit vector perpendicular to the surface.

On the other hand, the change of the number of particles inside the sphere with the radius R can be calculated by

$$\frac{dn_\alpha}{dt} = - 4 \pi R^2 j_R \qquad (5.13)$$

or if the cluster phase can be considered as incompressible (c_α = const) by

$$\frac{dR}{dt} = - \frac{j_R}{c_\alpha} \qquad (5.14)$$

If in addition it is assumed that at every moment the cluster is surrounded by a quasi-stationary concentration profile given by eq. (5.11) we get instead of (5.14)

$$\frac{dR}{dt} = \frac{D}{c_\alpha} \frac{c_\infty - c_R}{R} \qquad (5.15)$$

This equation describes the so-called diffusion limited growth, since the growth of the cluster is limited by the process of diffusion of the particles to the cluster.

For spherical clusters it can be shown, that the assumption of a quasi-stationary profile around the growing cluster is a good approximation. c_R is determined hereby based on the assumption of a local equilibrium between the cluster and the surrounding medium at the boundaries of the cluster.

If not the diffusion but processes at the interface between both phases determine the rate of growth, R in eq. (5.15) has to be replaced by a constant l_0. This type of growth is called interface kinetic limited growth and is described by

$$\frac{dR}{dt} = \frac{D}{c_\alpha} \frac{c_\infty - c_R}{l_0} \qquad (5.16)$$

Both equations (5.15) and (5.16) have a number of different applications (see, e.g., TAMMANN, 1922; RAWSON, 1967; MATZ, 1969) but problems arise, if additional factors, e.g., elastic fields influence the growth of the clusters.

In the next chapter we derive a general growth equation which allows without principal difficulties to take into account also such effects and to obtain growth equations for a variety of real systems.

5.3.2. Derivation of a General Growth Equation for Clusters of a New Phase

If we consider systems divided by boundaries between different phases, eq. (5.5) can be written as a difference equation of the form

$$\vec{J}_R = - \frac{Dc}{kT} \frac{\Delta\mu}{l} \vec{e}_r \tag{5.17}$$

$\Delta\mu$ is now the difference between the chemical potential in the both nearly homogeneous phases and l is the width of the inhomogeneous region between them.

In general, the thermodynamic driving force for the growth of the cluster is not given only by the difference of the chemical potentials but is influenced also by surface effects, elastic strains and other factors.

From thermodynamics it is known, that processes take place in the system, if the characteristic thermodynamic potential (F,U,H,G) decreases. So, as the thermodynamic driving force of the growth of a cluster we can consider the change of the thermodynamic potential $\Delta\Phi$ connected with an increase of the number of particles Δn_α in the cluster. Since the driving force may depend on the size of the cluster, we go over to the limit $\Delta n_\alpha \rightarrow 0$ and replace $\Delta\mu$ by

$$\Delta\mu \rightarrow - \frac{\Delta\Phi}{\Delta n_\alpha} \rightarrow - \frac{\partial\Phi}{\partial n_\alpha} \tag{5.18}$$

resulting in the following expression for j_R:

$$j_R = \frac{Dc}{kT} \frac{1}{l} \frac{\partial\Phi}{\partial n_\alpha} \tag{5.19}$$

The change of the number of particles in the cluster is given according to eq. (5.13) then by

$$\frac{dn_\alpha}{dt} = - 4\pi R^2 \frac{Dc}{kT} \frac{1}{l} \frac{\partial\Phi}{\partial n_\alpha} \tag{5.20}$$

and assuming incompressibility of the cluster phase by

$$\frac{dR}{dt} = - \frac{Dc}{c_\alpha kT} \frac{1}{l} \frac{\partial\Phi}{\partial n_\alpha} \tag{5.21}$$

For allotropic transitions or interface kinetic limited growth the width of the inhomogeneous region between both phases is equal to a constant $l = l_0$. The value of l for diffusion limited growth is determined based on the limiting condition that for ideal solutions eq. (5.15) has to be obtained, again.

For simplicity let us assume, that the cluster is formed by one of the components of the solution only. Due to the results presented in Chapter 3 the change of the Gibbs free energy can be expressed by

$$\Delta G = n_\alpha(\mu_\alpha(p) - \mu(p,x)) + \sigma A \qquad (5.22)$$

Substitution of ΔG into eq. (5.21) results then in

$$\frac{dR}{dt} = \frac{Dc}{c_\alpha kT} \frac{1}{l} \left\{ \mu(p,c) - \mu_\alpha(p) - \frac{2\sigma}{c_\alpha R} \right\} \qquad (5.23)$$

c is always equal to c_∞ here.
Neglecting as in eq. (5.15) surface effects ($\sigma = 0$) and taking into account the condition of the local equilibrium at the interface between cluster and medium ($\mu_\alpha = \mu(p,c_R)$) we get

$$\frac{dR}{dt} = \frac{Dc}{c_\alpha kT} \frac{1}{l} \left\{ \mu(p,c) - \mu(p,c_R) \right\} \qquad (5.24)$$

Applying further eq. (5.6) (ideal mixture) it follows

$$\mu(c_R) - \mu(c) = kT \ln \frac{c_R}{c_\infty} \approx kT \frac{c_R - c_\infty}{c_\infty} \qquad (5.25)$$

By substitution into eq. (5.24) one obtains

$$\frac{dR}{dt} = \frac{D}{c_\alpha} \frac{c_\infty - c_R}{l} \qquad (5.26)$$

A comparison with eq. (5.15) indicates that in the case of diffusion limited growth l is equal to the radius of the cluster.
Moreover, eq. (5.23) leads also to the conclusion, that the kinetic description is in agreement with the thermodynamic analysis. There exists a critical cluster radius R_c given by

$$\frac{dR}{dt} = 0 \qquad R_c = \frac{2\sigma}{c_\alpha(\mu(p,c) - \mu_\alpha(p))} \qquad (5.27)$$

A substitution of the expression for the critical radius into eq. (5.23) results in the known equation:

$$\frac{dR}{dt} = \frac{2D\sigma c}{c_\alpha^2 kT} \frac{1}{l} \left\{ \frac{1}{R_c} - \frac{1}{R} \right\} \qquad (5.28)$$

Eq. (5.28) is closely related to the deterministic growth equation of the cluster discussed in Chapter 4.5 . Again, for $R < R_c$ the cluster shrinks and disappears, for $R > R_c$ the cluster grows.
The application of the growth equations to different special problems is discussed in the Chapters 7 and 8. In Chapter 6 these equations are used first to give a description of the process of Ostwald ripening,

the competitive growth of an ensemble of clusters in the third stage
of a first-order phase transition according to the general scenario
outlined in Chapter 5.1.

5.4. Simultaneous Description of Nucleation and Growth of Clusters

Experiments, carried out by SUNDQUIST, ORIANI (1962), HEADY, CAHN
(1973), GOLDBURG, HUANG (1975) and others, on nucleation in the vicinity of the critical point showed, that intensive nucleation was re-
gistered only for higher supersaturations as predicted by nucleation
theories. These results were considered first as a refutation of the
classical and equivalent nucleation theories.
Binder and Stauffer, however, draw the attention to the fact, that in
general in experiments not nucleation itself but the result of nucleation and simultaneous growth of the already formed supercritical clusters is observed. Consequently, the deviation of the measurements
from the theoretical predictions can also be interpreted as the result
of the decrease of the diffusion coefficient and, therefore, of the
growth rate of the clusters in the vicinity of the critical point.

Binder and Stauffer developed a theory for the description of simultaneous nucleation and growth. They assumed that the nucleation rate
is constant and neglected the critical radius compared with the radius
of the growing clusters. As one result of the theory they determined
the so-called completion time, defined as the time-interval from the
beginning of the transition to the moment, when the actual supersaturation is equal to one half of the initial supersaturation (BINDER,
STAUFFER, 1976).
LANGER and SCHWARTZ (1980) proposed an improved version of a theory of
coupled nucleation and growth. In the limit of small supersaturations
(for which the theory is strictly valid only) both approaches are equivalent. Though both theories are not totally satisfying, the main
idea - the necessity of a simultaneous consideration both of nucleation and growth - is very important.

Confirming this idea we present computer calculations which model the
whole process of the phase transition. To consider the simultaneous
nucleation and growth of the clusters we now combine the quasi-steady-
state approximation of the nucleation rate (Chapter 5.2.) with the
growth equation of the cluster (Chapter 5.3).
The number of supercritical clusters formed in a time intervall Δt

can be calculated with the quasi-steady-state nucleation rate according to the eqs. (5.1) - (5.3). It is assumed that the clusters in the given time intervall are formed with the same radius $r_\alpha = 1.1\, r_c(\varrho_\beta)$. $r_c(\varrho_\beta)$ (eq. (5.3)) is the critical radius at the given moment t. Because ϱ_β decreases with time due to the formation and growth of the clusters (compare eq. (5.3)), the critical radius increases with time. The further evolution of the supercritical drops is described by the deterministic equation derived before:

$$\frac{dr_\alpha}{dt} = \frac{2\,\sigma\, D\,\varrho_\beta}{\varrho_\alpha^2\, R_g T}\,\frac{1}{r_\alpha}\left\{\frac{1}{r_c} - \frac{1}{r_\alpha}\right\} \tag{5.29}$$

with ϱ_β and r_c given by eq. (5.3).
In Fig. 5.2 the time evolution of the ensemble of drops is shown as a function of time for the condensation of water vapor in an one-component isochoric system.
We use a characteristic time τ determined by

$$\tau = \frac{27\,\varrho_\alpha^2\, R_g T\, r_{co}^3}{16\,\sigma\,\varrho\, D} \tag{5.30}$$

ϱ is the density of the initial vapor state and r_{co} is the radius of the critical cluster for $\varrho_\beta = \varrho$.
The time intervall Δt is chosen in such a way that after $10\,\Delta t$ the process of nucleation is practically finished.
For the chosen values of the parameters the period of active nucleation is equal to $2844\,\tau$ (or $2.9 \cdot 10^{-7}$s). The rapid decrease of the nucleation rate is due both to the formation of new and the growth of the already formed clusters.
Since the growth of the clusters contributes considerably to the decrease of the initial supersaturation, the number of clusters formed in the nucleation period is $1.6 \cdot 10^{15}\ \mathrm{m}^{-3}$, that is much less compared with the number of drops formed in an intervall of 10 s ($2.3 \cdot 10^{22}\mathrm{m}^{-3}$) if a possible simultaneous growth is neglected.
The critical size of the clusters in this first period increases only insignificantly ($r_c \lesssim 1.15\, r_{co}$), but due to the exponential dependence of the nucleation rate on r_c^2 it leads to significant decrease of the nucleation rate by a factor 10^{-1}.
In agreement with the general scenario given in Chapter 5.1. the second stage of the phase transition is characterized by an independent growth of the mean radius of the ensemble of clusters, their number being nearly constant. This stage can be divided into two different periods. In the first period the drops grow due to the consumption of monomers and the supersaturation decreases rapidly. For a time $t_c =$

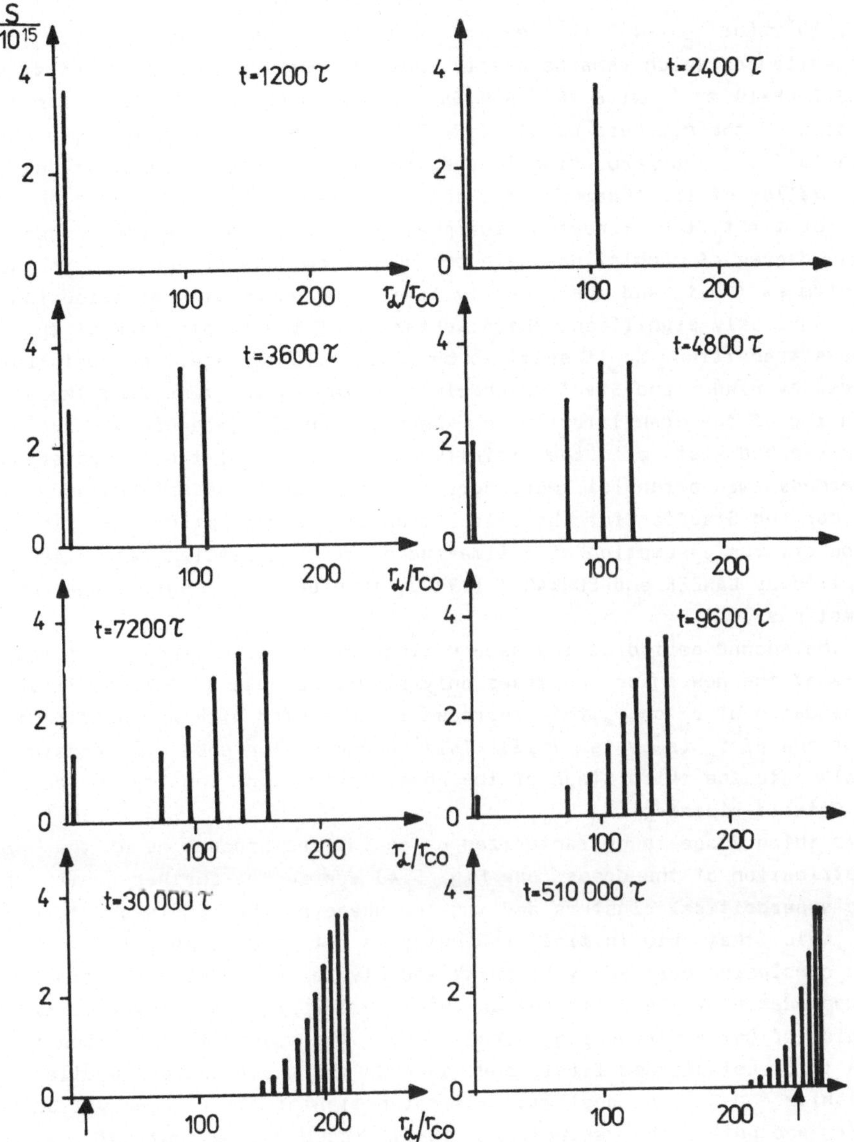

Fig. 5.2:
Formation and evolution of the droplet size distribution in the first stages of the phase transition. The arrow indicates the actual value of the critical radius.
Vapour: water, T = 293.15 K, initial supersaturation y_0 = 8
For the given values of the parameters it holds:

r_{co} = 5.24·10^{-10} m; τ = 1.03·10^{-10} s.

$4.1 \cdot 10^4 \tau$ (or $t_c = 4.2 \ 10^{-6}$ s) a size distribution of the drops is established, which remains nearly constant for longer periods of time $t \lesssim 1.48 \cdot 10^5 \tau$ ($t \lesssim 1.6 \cdot 10^{-5}$ s). This state is characterized by a mean radius of the clusters equal to $R = 246 \ r_{co}$ and a number of drops s = $1.6 \cdot 10^{15} \ m^{-3}$, corresponding to a state in the immediate vicinity of the valley of the thermodynamic potential (see also Fig. 5.3).

For $t = t_c$ the process of the phase transition has reached a certain degree of completion. Both the number of clusters present in the system at $t = t_c$ and their nearly time-independent size distribution are obviously significant characteristics of the first stage of the phase transition. So it seems to be reasonable to redefine the introduced by Binder and Stauffer completion time as the time from the beginning of the transition to the moment, when the ensemble of clusters has reached a state in the vicinity of the valley of the characteristic thermodynamic potential. Moreover, the additional assumptions used by Binder and Stauffer for the calculation of the completion time, in particular, the assumption of a time-independent nucleation rate, also applied by LANGER and SCHWARTZ (1980), turn out to be only crude approximations.

In the second period of the second stage of the transition the total mass of the new phase increases only slowly but due to the sensitive dependence of r_c on ϱ_ρ this increase is connected with a significant increase of r_c (see Fig. 7.15). This second period goes over continuously into the third stage of the phase transition, into the process of Ostwald ripening.

This third stage is characterized by an initial broadening of the size distribution of the drops (see Fig. 5.4) due to the further growth of the supercritical clusters and the shrinkage of the clusters with $r_\alpha < r_c$. Note, that this initial broadening is not in contradiction with the conclusion derived by Lifshitz and Slyozov concerning the time-independence of the distribution function $f(r_\alpha / r_c)$ in the asymptotic region of Ostwald ripening, since, first, the asymptotic distribution has to be established first, and, second, the distribution function, Lifshitz and Slyozov deal with, is a function of the reduced variables r_α / r_c and not of the variable r_α itself (compare, in contrast, KRISHNAMURTHY, GOLDBURG, 1980).

We finally remark that the characteristic time of duration of the different stages differ considerable ($t_1 < t_2 < t_3$), they decrease with an increasing initial supersaturation.

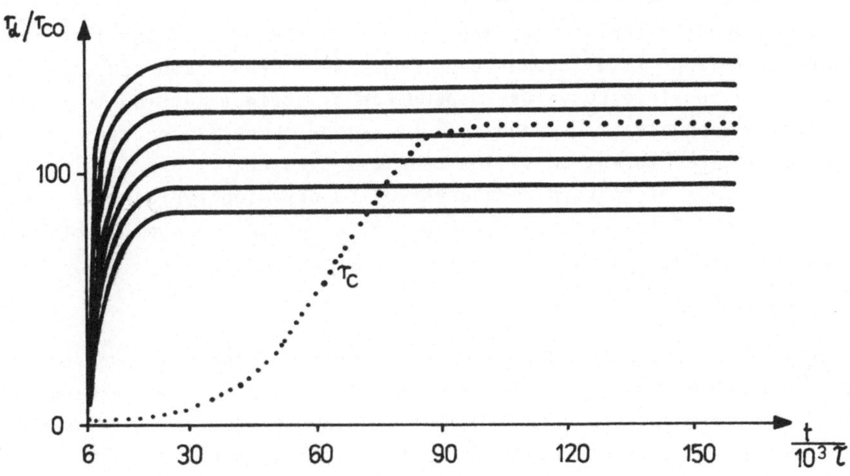

Fig. 5.3:
Evolution of the ensemble of clusters and the critical radius (dotted curve) in the stage of independent growth ($y_0 = 10$)

(From: SCHMELZER, 1985a)

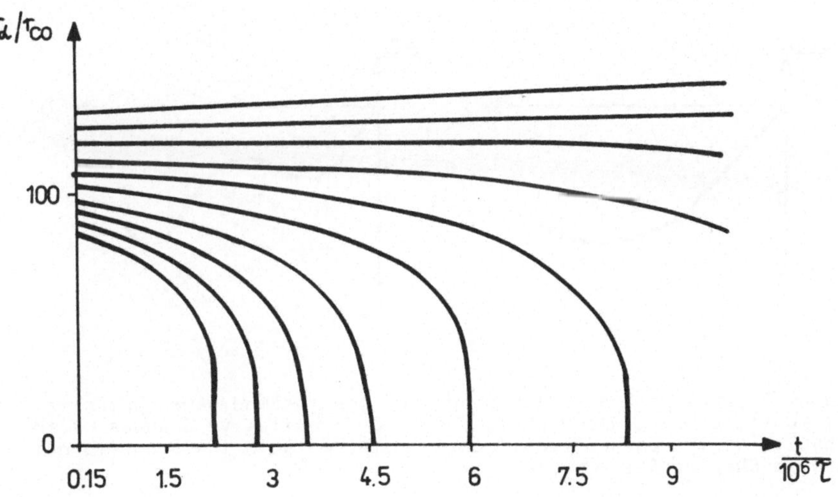

Fig. 5.4:
Evolution of the ensemble of drops in the stage of Ostwald ripening ($y_0 = 10$)

(From: SCHMELZER, 1985a)

5.5. Curvature Dependence of Surface Tension and the Scenario of First-Order Phase Transitions

In the preceding analysis of the process of the phase transition the capillarity approximation was used. Now, as it was discussed in Chapter 2.3. a possible curvature dependence of surface tension is widely discussed in nucleation theory as one method to obtain correct results for the work of formation of the critical clusters.
If the surface of tension is chosen as the dividing surface the work of formation of the critical clusters also for the case of a curvature dependent surface tension is given by

$$W = \frac{1}{3}\, \sigma\, A \tag{5.31}$$

(the more general case of an arbitrary choice of the dividing surface is discussed by PARLANGE, 1969; SCHMELZER, 1985b). It follows immediately, that a curvature dependence of surface tension may lead to significant quantitative variations of the nucleation rate.
Here another problem is of interest: Can a curvature dependence of surface tension in the region of applicability of the cluster model lead to qualitative changes of the mechanism of the phase transition?

Obviously, such a qualitative change is possible only, if the change of the thermodynamic potential instead of the behaviour shown in Fig. 3.1 is described by the curves shown in Fig. 5.5.

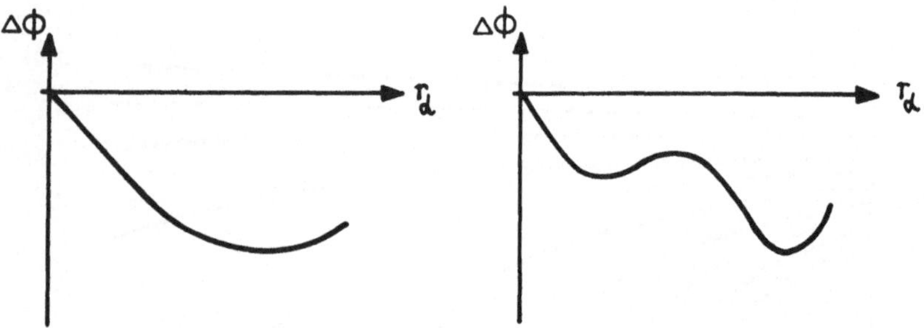

Fig. 5.5:
Hypothetic dependences of the change of the thermodynamic potential as a function of r_α which would result in qualitative changes of the mechanism of the phase transition as compared with the scenario outlined in Chapter 5.1.

Since the curvature corrections to the surface tension are of importance only for a small size of the cluster, it is necessary to analyze first the behaviour of $\Delta\Phi$ in the limit $r_\alpha \rightarrow 0$ and to check, if the inequality

$$\lim_{r_\alpha \to 0} \frac{\partial \Delta \phi}{\partial r_\alpha} < 0 \qquad\qquad (5.32)$$

can be fulfilled. The limit is to be understood in a physical sense as
the limit for very small but physically reasonable values of r_α .

Here we restrict ourselves to the investigation of the formation of
clusters with a higher compared with the medium density in one-component closed isochoric systems under isothermal conditions. The change
of the Helmholtz free energy due to the formation of a cluster can be
expressed then by eq. (3.5). Assuming, again, that at each moment the
mechanical equilibrium is practically established the derivative of
ΔF with respect to r_α reads:

$$\frac{\partial \Delta F}{\partial r_\alpha} = - 4\pi r_\alpha^2 \left\{ p_\alpha - p_\beta - \frac{2\sigma}{r_\alpha} - \frac{\partial \sigma}{\partial r_\alpha} - (\mu_\alpha - \mu_\beta) \frac{dn_\alpha}{dV_\alpha} \right\} \qquad (5.33)$$

By a Taylor expansion of μ_α and μ_β , taking into account both the conditions of a mechanical equilibrium and $\varrho_\alpha \gg \varrho_\beta$ we obtain approximately

$$\frac{\partial \Delta F}{\partial r_\alpha} \simeq - 4\pi r_\alpha^2 \left\{ \frac{dn}{dV} \frac{p_\beta - p'}{\varrho_\beta(p')} - \frac{\varrho_\alpha(p_\alpha)}{\varrho_\alpha(p')} \frac{2\sigma}{r_\alpha} - \frac{\partial \sigma}{\partial r_\alpha} \right\} \qquad (5.34)$$

The first term in the brackets in eq. (5.34) represents the thermodynamic driving force of the phase transition. It has positive finite
values.
Applying the capillarity approximation for small values of r_α always
the second term B is dominating.

$$B = - \frac{\varrho_\alpha(p_\alpha)}{\varrho_\alpha(p')} \frac{2\sigma}{r_\alpha} - \frac{\partial \sigma}{\partial r} \qquad\qquad (5.35)$$

So the question is to be analyzed, whether for special types of the
curvature dependence of surface tension the second term can have a
smaller absolute value compared with the first.
This analysis is carried out here in detail for equations of the type
(2.40), the conclusions remain valid also for a dependence of the
surface tension on curvature determined by eq. (2.39).
First of all it can be noticed, that all values of the parameter $a < 0$
have to be excluded, since for $r_\alpha \to |a|, \delta \to \infty$. δ can tend to infinity
only if $\varrho_\alpha \to \varrho_\beta$ (see eq. (2.33)), but in this case the cluster model is
not applicable.
In the case $a > 0$ B always tends to infinity. Only for $a = 0$ finite
negative values of B are obtained in the limit $r_\alpha \to 0$.

But for the considered case the critical radius (using the capillarity
approximation) is given by

$$r_c \simeq \frac{2\sigma\,\varrho_\beta(p')}{\varrho_\alpha(p')(p_\beta-p')} \qquad (5.36)$$

and eq. (5.34) yields

$$\lim_{r_\alpha\to 0}\frac{\partial\Delta F}{\partial r_\alpha} \simeq - 4\pi r_\alpha^2 \frac{(p_\beta-p')\varrho_\alpha(p')}{\varrho_\beta(p')}\left\{\frac{1}{\varrho_\alpha(p')}\frac{dn_\alpha}{dV_\alpha} - \frac{r_c}{\delta_\infty}\left[\frac{\varrho_\alpha(p)}{2\,\varrho_\alpha(p')} + \frac{1}{4}\right]\right\} \qquad (5.37)$$

In the limit of applicability of thermodynamics r_c as the radius of
the critical cluster must be considerably greater than δ_∞ being smaller
than the width of the interfacial region. Therefore, also for Tolman's
equation qualitatively there is no change of the scenario of the phase
transition.

We come, consequently, to the conclusion, that in the region of applicability of the cluster concept, a curvature dependence of the surface
tension cannot lead to a qualitatively different scenario of the phase
transition.

On the other hand, a situation as expressed in Fig. 5.5 corresponds
to the process of spinodal decomposition. As it was already mentioned,
in this case the transition proceeds via the formation of long wave-
length fluctuations with initially small differences in the density.
We came here to the same conclusions based on a thermodynamic approach. This confirms, to our opinion, in addition the proposed by us
equations for the curvature dependence of surface tension.

5.6. Further Applications

In the present chapter, in general, nucleation and growth processes
under isothermal conditions were considered. In experimental or practical situations other constraints are also of importance, in particular, adiabatic conditions.

For adiabatic conditions the thermodynamic analysis of nucleation and
growth can be carried out in the same way as outlined here with qualitatively analogous results (SCHMELZER, 1984a; SCHMELZER, SCHWEITZER,
1986). The mechanism of growth may, however, change e.g. from kinetic
limited growth or from limitation by the diffusion of particles to limitation by heat conduction.

The general feature of all cases analyzed was, that for the chosen
constraints either no or only one stable heterogeneous state - cluster

in the otherwise homogeneous medium - exists: In multicomponent systems according to Gibbs' phase rule heterogeneous systems consisting of more than two phases with different compositions are possible. Consequently, in these more general cases more than one valley of the thermodynamic potential may exist, corresponding to different possibilities of formation and evolution of the new phase (see also NEUMANN, DÖRING, 1940).
The developed method of kinetic description of the growth of the new phase is also applicable, if simultaneously to the growth of the new phase chemical reactions take place (see also KATZ, DONOHUE, 1982).

Many areas of current interest in the material sciences involve the simultaneous diffusional growth and dissolution of a randomly dispersed second phase. This competitive growth problem (Ostwald ripening) will be discussed in detail in Chapter 6. Here we finally present deterministic equations of motion, which are based on a semi-phenomenological model of the growth of AgCl nuclei in a model photochromic glass melt (BUDDE, PASCOVA et al., 1986).
The equations for the temporal change of the volume $V_\alpha = (4\pi/3)\, r_\alpha^3$ of the spherical nuclei read as follows:

$$\frac{dV_\alpha}{dt} = 4\pi D\, c'\; \frac{2\sigma}{\rho_\alpha R_g T}\; \left\{ (r_\alpha/r_c) - 1 \right\} - (V_\alpha - V_0)/\tau_{rel} \qquad (5.38)$$

with

$$r_c = \frac{2\sigma}{\rho_\alpha R_g T\, \ln(c_\beta/c')} \qquad (5.39)$$

and

$$c_\beta(t) = c_0 - c_\infty(1 - e^{-kT}) \quad . \qquad (5.40)$$

The deterministic growth equation (5.38) consists of the well-known growth law and an additional contribution related to relaxation processes in the matrix. The quantities V_0 (initial volume) and τ_{rel} (relaxation time) are experimental parameters. The critical radius r_c (5.39, compare (3.28)) is changing by depletion of the dissolved matrix particles. Here we use the empirical expression (5.40) with the parameters c_0, c_∞ (concentrations) and k (reciprocal time unit), which makes the set of equations complete. Numerical integration of eq. (5.38) with eqs. (5.39, 5.40) gives the volume as function of time $V_\alpha(t)$. In dependence of the relaxation time τ_{rel} solutions are

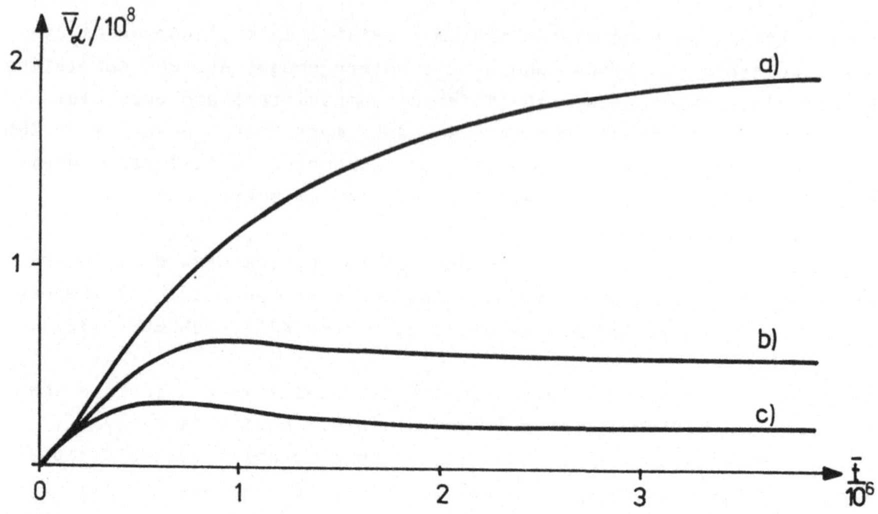

Fig. 5.6:
Temporal change of the reduced volume $\overline{V}_\alpha = V_\alpha/\Omega$ in dependence of the dimensionless time $\overline{t} = t/\tau$ for different relaxation times τ_{rel}, expressed by $\tau_{rel}^{-1} = \tau/\tau_{rel}$ ($\Omega = 6.24\ 10^{-3}\ nm^3$, $\tau = 4.46\ 10^{-4}\ s$)
a) $\tau_{rel}^{-1} = 1\cdot10^{-6}$ b) $\tau_{rel}^{-1} = 2.5\cdot10^{-6}$ c) $\tau_{rel}^{-1} = 5\cdot10^{-6}$
(From: BUDDE et al., 1986)

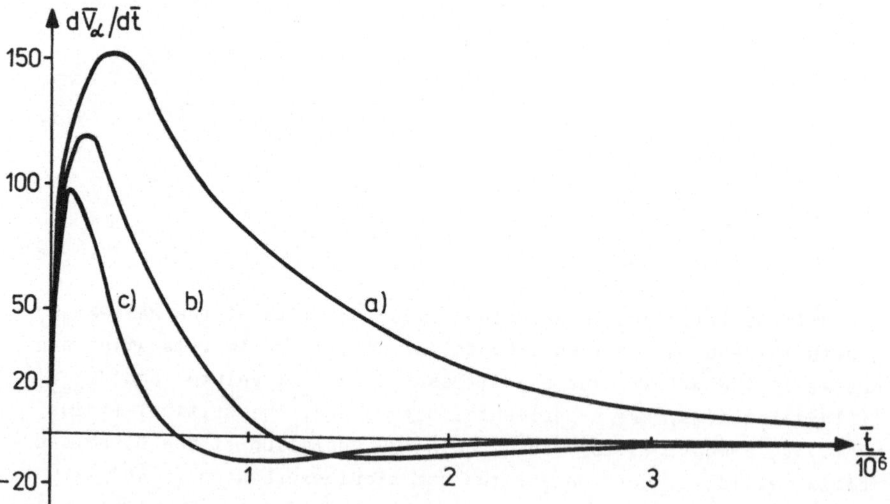

Fig. 5.7:
Time dependence of the velocity $\dot{\overline{V}}_\alpha = d\overline{V}_\alpha/d\overline{t}$ for the same parameters τ_{rel}^{-1} given above.
(From: BUDDE et al., 1986)

shown in Figs. 5.6, 5.7 using reduced quantities. After about thirty
minutes (t = $\tau \cdot \bar{t} \approx 30$ min) the growth of the AgCl nuclei has finished
and reached a value of $V_\alpha \approx 10^6$ nm^3. These results are related to
experiments (BUDDE et al., 1986; PASCOVA, GUTZOW, 1983). The saturation effect can be also explained by the influence of elastic or
viscoelastic forces (see Chapter 8).

It was shown, that the general scenario of the phase transition is
a result of the depletion effects and widely independent of any specific properties of the system considered. Consequently, the same general scenario is expected to be valid under certain conditions for
phase transitions in two-dimensional systems, e.g., adsorbed mono-
layers (GEBRE MEDHIN, SCHMELZER, 1987). The application of the theoretical concepts outlined to the mentioned additional applications is
a topic of current research.

6. Theory of Ostwald Ripening

6.1. Basic Equations

Since the process of coarsening, the growth of the mean radius of the
clusters of a new phase in solutions connected with a decrease of their
number was described and discussed first by OSTWALD (1901), such a
process is widely denoted as Ostwald ripening (LIESEGANG, 1911).

Attempts to a theoretical explanation of Ostwald ripening were pro-
posed by a number of authors (TODES, 1946, 1947; GREENWOOD, 1956;
LIFSHITZ, SLYOZOV, 1958, 1959, 1961; WAGNER, 1961; KAHLWEIT, 1970,
1975).Developments in the theory of Ostwald ripening since the classic
work by Lifshitz and Slyozov are summarized and extended recently by
Marqusee, Ross, Venzl, Voorhees and others (MARQUSEE, ROSS, 1983, 1984;
MARQUSEE, 1984; BEENAKKER, ROSS, 1985; VENZL, 1983, 1985; VOORHEES,
GLICKSMAN, 1984; VOORHEES, 1985; TOKUYAMA et al., 1984, 1986; KAWASAKI,
ENOMOTO, TOKUYAMA, 1986). Direct relations of the process of Ostwald
ripening to the theory of selforganizing systems are underlined by
MAHNKE, FEISTEL (1985), MAHNKE (1987) and will be discussed in detail
in Chapter 6.5.

In general the time evolution of first order phase transitions can be
divided in three stages, which are characterized as follows:
i)   Nucleation
     New phase (α) is formed out of the metastable old phase (β); large
     number of clusters with a narrow distribution of sizes
ii) Growth
     Nuclei grow rapidly at the expense of the old phase; at the end
     the mass of the new phase is close to its equilibrium value
iii) Coarsening or Ostwald ripening
     New phase evolves so as to minimize its energy; drastic reduction
     in the number of particles of the new phase; larger particles grow
     at the expense of smaller ones, which disappear.

The theoretical description of the process of phase transitions can be
considered at different levels. Apart from the concept discussed in
this booklet, let us mention that the time evolution of phase separation can be described by means of a concentration field $c(r,t)$ depending on the coordinate vector r and time t. This local concentration is
the basic quantity, which has to fulfil a reaction-diffusion equation
(SCHIMANSKY-GEIER, EBELING, 1983; MALCHOW, SCHIMANSKY-GEIER, 1985;
FEISTEL, 1986). Depending on the boundary conditions the growth of do-
mains (phases with nearly constant concentration) are studied

144

(SEKIMOTO, 1986). The theoretical understanding of domain-growth kinetics in quenched systems undergoing ordering processes is of enormous
interest (MILCHEV, BINDER, HEERMANN, 1986). The simplest model of this
so-called nonequilibrium phase transition is a bistable chemical reaction system (Schlögl model), which shows spatial separation of phases
in droplet-like configurations. More complicated situations of spontaneous pattern formation in precipitating systems are explained by
FEINN et al., 1978 and LOVETT, ORTOLEVA, ROSS, 1978 for the Liesegang
experiments.

In our approach we consider a model where spherical droplets are located in the otherwise homogeneous vapour. The system of volume V is
well stirred, so we do not consider the coordinates of the clusters.
Let us introduce the cluster density distribution function $f(n,t)$ where
n is the number of bounded in the cluster monomers (n ~ cluster size)
and $f(n,t)dn$ the number of clusters per volume with a size between n
and n+dn. Now the basic quantities describing the system are

$$f(n,t) = \frac{1}{V} \frac{dN_n(t)}{dn} \qquad \text{cluster distribution density} \qquad (6.1)$$

$$c_\beta(t) = N_1(t)/V \qquad \text{concentration of free monomers} \qquad (6.2)$$

where $N_1(t)$ is the number of free monomers in the volume V of the system and $N_n(t)$ the number of droplets with n bounded particles respectively. The quantities (6.1, 6.2) are the continuous analogies to the
discrete cluster distribution $\underline{N}$ introduced in Chapter 4 by eq.(4.3).
   We mention (ULBRICHT et al., 1986; MAHNKE, 1986) that the distribution density (6.1) is normalized to the total droplet density (N_α -
number of all drops)

$$\int_0^\infty f(n,t)\ dn = N_\alpha/V \qquad (6.3)$$

and for the first moment of the distribution function we get the total
number of bounded monomers M_α per volume

$$\int_0^\infty n\ f(n,t)\ dn = M_\alpha(t)/V \quad . \qquad (6.4)$$

In the following, we formulate the basic equations of the competitive
growth problem. In the space of cluster sizes the distribution density
$f(n,t)$ (eq. (6.1)) satisfies a balance equation of the form

$$\frac{\partial f(n,t)}{\partial t} + \frac{\partial}{\partial n} (v\ f(n,t)) = j(t) \qquad (6.5)$$

with a growth rate of a droplet of size n in the presence of monomer concentration c_β (eq. (6.2))

$$\frac{dn}{dt} = v(n, c_\beta)$$ (6.6)

and a production-decay term j. In a finite system the quantities c_β and $f(n,t)$ are not independent from each other, they are coupled by conservation of matter

$$c_\beta(t) + \int_0^\infty n\ f(n,t)\ dn = N_m/V$$ (6.7)

where N_m (eq. (4.4)) the total number of particles (either free or bounded monomers) is fixed.

The production rate j in eq. (6.5) vanishes when nucleation or such phenomena like coagulation, splitting, etc., which introduce new droplets of a given size class, are negligible. That's why in the Ostwald ripening regime eq. (6.5) can be approximated by the continuity equation

$$\frac{\partial f(n,t)}{\partial t} + \frac{\partial}{\partial n}\ (v\ f(n,t)) = 0$$ (6.8)

which is a restricted form of the Fokker-Planck equation (see Chapter 4.5.). A comparison of eq. (6.8) with eq. (4.88) shows how the cluster density is related to the mean values of the number of clusters $\langle N_n \rangle$. It is obvious (RÖPKE, 1987; SUBAREW, 1976) that the Liouville equation (6.8) describes the evolution of the cluster distribution in a deterministic sense. We derive deterministic equations for the competitive growth of an ensemble of droplets in Chapter 6.5.

A number of authors have been discussed the problem of Ostwald ripening. An overview of the theory of Ostwald ripening or coarsening is given by VOORHEES (1985). Numerical solutions of eqs. (6.6-6.8), written in dimensionless form, are obtained by VENZL (1983), experimental results by HOHMANN (1974). But all these attempts are based on the classical work by Lifshitz and Slyozov. Because of its importance in the next Chapter the main assumptions and results of the Lifshitz-Slyozov theory are outlined.

6.2. The Lifshitz-Slyozov Theory

The kinetic description of coalescence phenomena given by ZELDOWITSCH (1942) is based on a kinetic equation of the Fokker-Planck type. Following LANDAU, LIFSHITZ (1983)   eq. (6.5) is written as

$$\frac{\partial f}{\partial t} = - \frac{\partial I}{\partial n} \qquad (6.9)$$

with

$$I = v\, f - D\, \frac{\partial f}{\partial n} \quad . \qquad (6.10)$$

In the stationary case ($\partial f/\partial t = 0$) the current I is constant and determines the stationary solution f^s. This procedure was already described in Chapter 1.3.2., starting with eq. (1.9) in analogy to eq. (6.9), and in Chapter 4.5.

LIFSHITZ and SLYOZOV (1958, 1959, 1961) started in their investigation of the process of Ostwald ripening (see also LANDAU,LIFSHITZ, 1983) with eq. (6.11) (compare eq. (4.94) or eq. (5.28) for similar growth laws)

$$\frac{dr_\alpha}{dt} = \frac{2\sigma c'D}{c_\alpha^2 kT} \frac{1}{r_\alpha} \left[\frac{1}{r_c} - \frac{1}{r_\alpha} \right] \qquad (6.11)$$

where the radius of the cluster is denoted by r_α. In comparison with eq. (5.23) the concentration in the surrounding the clusters medium is replaced by the constant equilibrium concentration c'. This approximation is accurate in the asymptotic stage of Ostwald ripening for which solutions are obtained. Introducing the notations

$$\alpha' = \frac{2\sigma c'}{c_\alpha^2 kT} \qquad t' = \frac{t}{T} \qquad T = \frac{r_{co}^3}{\alpha' D} \qquad \wp = \frac{r_\alpha}{r_{co}} \qquad (6.12)$$

r_{co} being the critical radius for t = 0, eq. (6.11) can be written as

$$\frac{d\wp^3}{dt'} = 3\left(\frac{\wp}{x} - 1\right) \qquad \text{with} \quad x = \frac{r_c}{r_{co}} \quad . \qquad (6.13)$$

The quantity x is an increasing function of time and tends to infinity for $t \rightarrow \infty$.

Lifshitz and Slyozov introduced also a function f describing the distribution of the clusters with respect to their sizes. For this function a continuity equation (6.8) is fulfilled

$$\frac{\partial f}{\partial t} + \frac{\partial}{\partial \wp} (f\, v_\wp) = 0 \qquad (6.14)$$

since in the considered stage of the phase transition the spontaneous formation of new clusters is excluded.

The conservation of the number of monomers results in the following equation

$$1 = \frac{c_\beta - c'}{c - c'} + \frac{4\pi}{3} \frac{c_\alpha r_{co}^3}{c - c'} \int_0^\infty \frac{f(\varrho)}{V} \, \varrho^3 d\varrho \tag{6.15}$$

c_β is the actual, c the initial and c' the equilibrium concentration of monomers of the segregating particles in the medium. Introducing further two new variables

$$\tau = \ln x^3(t') \qquad u = r_\alpha/r_c \tag{6.16}$$

Lifshitz and Slyozov arrived at

$$\frac{du^3}{d\tau} = \gamma(\tau) [u - 1] - u^3 \qquad \gamma(\tau) = \frac{3dt'}{dx^3} \quad . \tag{6.17}$$

Since r_c tends to infinity the condition of conservation of mass eq. (6.15) can be fulfilled only, if γ tends to 27/4 for long times. In this asymptotic region the mean radius of the ensemble of clusters behaves as

$$R^3 = \frac{8}{9} \frac{D\sigma}{kT} \frac{c'}{c_\alpha^2} t \tag{6.18}$$

and the number of clusters decreases as

$$s = V \frac{0.99}{4} \frac{c - c'}{c'} \frac{c_\alpha kT}{D\sigma} \frac{1}{t} \tag{6.19}$$

The mean radius of the ensemble of clusters in the asymptotic region is equal to the critical radius r_c. The asymptotic solutions according to Lifshitz and Slyozov are valid for times $t \gg kTc_\alpha^3 \, r_{co}^3/(2\sigma Dc')$, where r_{co} is not precisely defined. Moreover, it is shown, that the distribution function $f(r_\alpha/r_c)$ is time-independent in the asymptotic region.

The results of Lifshitz and Slyozov obtained from diffusion limited growth are verified both by a large number of experimental observations (e.g., JAIN, HUGHES, 1978) and computer simulations (e.g., PENROSE et al., 1978). Though the results are mathematically exact in the region of applicability of the used concepts (the number of clusters per unit volume must be sufficiently high, otherwise a continuous description in form of a distribution function in cluster sizes is not reasonable) there is some lack of insight into the physical background of the ripening process. Moreover, only asymptotic solutions are obtained and in the criteria of applicability of the asymptotic solutions the parameter r_{co} remains not precisely defined.

The disadvantages are removed in our approach we are going to discuss

now. Moreover, this new method has the additional advantages of mathematical simplicity, general applicability, e.g., the Ostwald ripening
of bubbles, Ostwald ripening under the influence of elastic strains.

The mathematical simplicity and general applicability are partly
due to the limitation (a disadvantage, of course), that from the very
beginning only the development of the mean cluster radius is considered. From the real distribution of clusters developing in the system
we go over to an idealized ensemble of s identical clusters and calculate the thermodynamic driving force of the evolution of the system
based on thermodynamic investigations of this idealized ensemble.
Therefore, based on this method, in general, no conclusions concerning
the distribution of clusters with respect to their sizes in the real
ensemble can be derived.

6.3. Thermodynamic Aspects of Ostwald Ripening in Solid and Liquid Solutions

We consider a binary liquid or solid solution under a constant external pressure p and a constant temperature T. The number of particles
of the different components is kept constant.
The change of the characteristic thermodynamic potential, in this case,
the Gibbs free energy, due to the formation of s clusters of the new
phase can be expressed, generally, as (see also: SCHMELZER, SCHWEITZER,
1987a,b,c; SCHMELZER, 1985a,c,d):

$$\Delta G = \sum_{j=1}^{s} \left\{ (p-p_\alpha^{(j)})V_\alpha^{(j)} + \sigma^{(j)}A^{(j)} + \sum_{i=1}^{k} (\mu_{i\alpha}^{(j)} - \mu_{i\beta}) \tilde{n}_{i\alpha} \right\} \\ + \sum_{i=1}^{k} (\mu_{i\beta} - \mu_i) n_i$$

(6.20)

In the considered case k is equal to two. It is assumed, further, that
$\tilde{n}_{i\alpha} = n_{i\alpha}$ (or $n_{io} = 0$), so that the surface tension depends only on
temperature. Moreover, the cluster is assumed to consist mainly of par-
ticles of the second component (solute) and is considered as incompressible. With this assumptions eq. (6.20) reads

$$\Delta G = \sum_{j=1}^{s} \left\{ (\mu_{2\alpha}^{(j)}(p,T) - \mu_{2\beta}(p,T,x_\beta))n_{2\alpha} + \sigma^{(j)}A^{(j)} \right\} + \\ + \sum_{i=1}^{2} (\mu_{i\beta} - \mu_i) n_i$$

(6.21)

Expressing the chemical potential of the matrix particles according to the perfect solution law by

$$\mu_i(p,T,x_i) = \mu_i(p,T) + kT \ln x_i \tag{6.22}$$

and taking into account the equilibrium condition $\mu_{2\alpha}(p,T) = \mu_{2\beta}(p,T,x')$ we get, finally,

$$\Delta G = \sum_{j=1}^{s} \left\{ -n_{2\alpha}^{(j)} kT \ln \frac{x_\beta}{x'} + \sigma^{(j)} A^{(j)} \right\} + n_2 \, kT \, \ln \frac{x_\beta}{x} + \\ + n_1 \, kT \, \ln \frac{1-x_\beta}{1-x} \tag{6.23}$$

If an ensemble of s identical clusters is considered this equation reads

$$\Delta G = - s n_\alpha kT \, \ln \left[\frac{c}{c'} \, \frac{(1-s\frac{n_\alpha}{n_2})}{(1-s\frac{V_\alpha}{V})} \right] + n_2 \, kT \, \ln \left[1 - \frac{s n_\alpha}{n_2} \right] \\ + n_1 \, kT \, \ln \left[\frac{1-x_\beta}{1-x} \right] + s \, \sigma \, A \tag{6.24}$$

For simplicity $n_{2\alpha}$ is written as n_α. The following other notations are used:

x,c,n - molar fraction, concentration and number of particles of the second component (solute) in the homogeneous initial state

x_β, c_β - variable molar fraction and concentration of the solute particles during the ripening process

x', c' - values of the molar fraction and concentration of the solute particles in the medium in equilibrium with the evolving α-phase.

Since we are dealing with solid or liquid solutions the volume V is nearly constant, moreover, $c_\alpha \gg c$ is assumed.

To allow later the consideration of the influence of elastic strains on the process of Ostwald ripening a term $\phi^{(\varepsilon)}(V_\alpha)$ is added to eq. (6.24) which describes the elastic effects. In this Chapter, after the general results are obtained, $\phi^{(\varepsilon)}$ is set equal to zero, again. Instead of eq. (6.24) eq. (6.25) is obtained now

$$\Delta G = - s n_\alpha kT \, \ln \left[\frac{c}{c'} \, \frac{(1-s\frac{n_\alpha}{n_2})}{(1-s\frac{V_\alpha}{V})} \right] + n_2 \, kT \, \ln \left[1 - \frac{s n_\alpha}{n_2} \right] \\ + n_1 \, kT \, \ln \left[\frac{1-x_\beta}{1-x} \right] + s \, \sigma \, A + \phi^{(\varepsilon)}(V_\alpha) \tag{6.25}$$

The extrema of ΔG are determined by

$$\left(\frac{\partial \Delta G}{\partial r}\right)_s = - 4\pi r_\alpha^2 c_\alpha kTs \left\{ \ln\left[\frac{c}{c'} \frac{(1-s\frac{n_\alpha}{n_2})}{(1-s\frac{V_\alpha}{V})}\right] + \frac{s\frac{V_\alpha}{V}}{1-s\frac{V_\alpha}{V}} \right.$$

$$\left. - \frac{2\sigma}{c_\alpha kTr_\alpha} - \frac{1}{4\pi r_\alpha^2 c_\alpha kT} \frac{\partial \phi^{(\varepsilon)}}{\partial r_\alpha} \right\} = 0 \qquad (6.26)$$

or by the generalized Gibbs-Thomson equation (6.27)

$$\ln\left[\frac{c}{c'} \frac{(1-s\frac{n_\alpha}{n_2})}{(1-s\frac{V_\alpha}{V})}\right] + \frac{s\frac{V_\alpha}{V}}{1-s\frac{V_\alpha}{V}} - \frac{2\sigma}{c_\alpha kTr_\alpha} - \frac{1}{4\pi r_\alpha^2 c_\alpha kT} \frac{\partial \phi^{(\varepsilon)}}{\partial r_\alpha} = 0 \qquad (6.27)$$

This equation gives for the extrema of ΔG a relation between the number of clusters and their mean radius. Considering the radius as a continuous function of s which is possible because of the high number of clusters from eq. (6.27) the following equation can be derived

$$\frac{dr_\alpha}{ds} = - \frac{r_\alpha}{3s} \frac{Z}{1+Z- \frac{r_\alpha^2}{2\sigma} \frac{\partial^2 \phi^{(\varepsilon)}}{\partial V_\alpha \partial r_\alpha}} \qquad (6.28)$$

$$Z = 3 \frac{c_\alpha kTr_\alpha}{2\sigma} \frac{sV_\alpha}{V} \left\{ - \frac{c_\alpha}{c(1- \frac{sn_\alpha}{n})} + \frac{1}{1-s\frac{V_\alpha}{V}} + \frac{1}{(1-s\frac{V_\alpha}{V})^2} \right\} .$$

The quantity Z has a negative value. Further information on Z can be obtained from the extremum conditions. Since the second partial derivative of ΔG with respect to r_α is given by

$$\left(\frac{\partial^2 \Delta G}{\partial r_\alpha^2}\right)_s = - 8\pi\sigma s \left\{ 1 + Z - \frac{r_\alpha^2}{2\sigma} \frac{\partial^2 \phi^{(\varepsilon)}}{\partial r_\alpha \partial V_\alpha} \right\} \qquad (6.29)$$

the following relations hold

$$1 + Z - \frac{r_\alpha^2}{2\sigma} \frac{\partial^2 \phi^{(\varepsilon)}}{\partial r_\alpha \partial V_\alpha} > 0; \ \frac{dr_\alpha}{ds} > 0 \quad \text{for } r_\alpha = r_c$$

$$\qquad (6.30)$$

$$1 + Z - \frac{r_\alpha^2}{2\sigma} \frac{\partial^2 \phi^{(\varepsilon)}}{\partial r_\alpha \partial V_\alpha} < 0; \ \frac{dr_\alpha}{ds} < 0 \quad \text{for } r_\alpha = r_{st}$$

The variation of the extreme values of ΔG with an increasing number of clusters can be described by eq. (6.31)

$$\frac{d \Delta G}{ds} = \frac{1}{3} \sigma A + \phi^{(\varepsilon)} - V_\alpha \frac{\partial \phi^{(\varepsilon)}}{\partial V_\alpha} \qquad (6.31)$$

If we consider now the case $\phi^{(\varepsilon)} = 0$ (no elastic strain), a comparison with the results obtained in the analysis of the formation of drops in

one-component closed isochoric systems outlined in Chapter 3.6, shows
faregoing analogies. The equations (3.72) - (3.76) are identical with
eqs. (6.28) - (6.31). Therefore, ΔG as a function of the radius and
the number of clusters shows a behaviour, again, as presented in Fig.
3.18 for ΔF.
This is the first example indicating, that the scenario of the phase
transition is determined mainly by depletion effects and not by the
special properties of the system considered. These special properties
are reflected only in the special form of expression of the quantity Z.

Based on the outlined here results of thermodynamic investigations
in the next Chapter the kinetics of growth of the ensemble of clusters
in the third stage of the phase transition is analyzed. This third
stage is indicated in Fig. 3.18 by a dashed curve. It may start at the
highest point P of the "valley" of the characteristic thermodynamic
potential corresponding to a mean cluster radius r_{cs} and a number of
clusters s_c determined by

$$\left(\frac{\partial \Delta G}{\partial r_\alpha}\right)_s = \left(\frac{\partial^2 \Delta G}{\partial r_\alpha^2}\right)_s = 0 \qquad Z = -1 \quad . \qquad (6.32)$$

6.4. A New Method of Kinetic Description of Ostwald Ripening

In accordance with eq. (5.21) the evolution of a cluster of the real
ensemble can be expressed as

$$\frac{dr_\alpha^{(j)}}{dt} = - \frac{Dc_\beta}{c_\alpha kT} \frac{1}{1^{(j)}} \frac{\partial \Delta G}{\partial n_\alpha^{(j)}} \quad . \qquad (6.33)$$

In the derivation of ΔG, given here by eq. (6.23), with respect to
$r_\alpha^{(j)}$ the radii of all other clusters, except the j-th one, have to be
held constant. As the result we obtain:

$$\frac{dr_\alpha^{(j)}}{dt} = \frac{Dc_\beta}{c_\alpha} \frac{1}{1^{(j)}} \left\{ \ln \left[\frac{c}{c'} \frac{(1 - \frac{n_\alpha + n_\alpha'}{n_2})}{(1 - \frac{V_\alpha + V_\alpha'}{V})} \right] + \frac{\frac{V_\alpha + V_\alpha'}{V}}{1 - \frac{V_\alpha + V_\alpha'}{V}} - \frac{2\sigma}{c_\alpha kT r_\alpha^{(j)}} \right\} \quad .$$

$$(6.34)$$

V_α' and n_α' are the volume and the number of particles of the additional
s-1 particles at the given moment t. Consequently, according to eq.
(6.34) the value of the radius $r_c^{(j)}$, obeying the necessary thermodynamic equilibrium conditions, depends on the number and the size of all
other clusters. This equilibrium value can be determined by

$$\ln\left[\frac{c}{c'}\,\frac{\left(1-\dfrac{n_\alpha+n'_\alpha}{n_2}\right)}{\left(1-\dfrac{V_\alpha+V'_\alpha}{V}\right)}\right] + \frac{\dfrac{V_\alpha+V'_\alpha}{V}}{1-\dfrac{V_\alpha+V'_\alpha}{V}} = \frac{2\sigma}{c_\alpha kT r_c^{(j)}} \quad . \tag{6.35}$$

A substitution into eq. (6.34) leads to an equation of the form
(5.28), again. Since, in general, $V_\alpha \ll V'_\alpha$ and $n_\alpha \ll n'_\alpha$ the values of
the critical radii $r_c^{(j)}$ are nearly the same for all clusters and we can
write, approximately,

$$\frac{dr_\alpha^{(j)}}{dt} = \frac{2\sigma c_\beta D}{c_\alpha^2 kT}\,\frac{1}{1^{(j)}}\left\{\frac{1}{r_c} - \frac{1}{r_\alpha^{(j)}}\right\} \quad . \tag{6.36}$$

Eq. (6.34) can be used for numerical investigations of the process of
growth, in general, and Ostwald ripening, in particular.
Here based on thermodynamic methods a theory is developed, describing
how the mean radius of the ensemble of clusters R and its number s
change with time. Since the number of clusters is assumed to be a big
one, in the derivation we may use the eqs. (6.28) – (6.31).
Multiplying eq. (6.33) with the number of clusters s we obtain

$$s\,\frac{dr_\alpha^{(j)}}{dt} = -\frac{Dc_\beta}{c_\alpha^2 kT}\,\frac{1}{1^{(j)}}\,\frac{s}{4\pi r_\alpha^{(j)2}}\,\frac{\partial\Delta G}{\partial r_\alpha^{(j)}} \tag{6.37}$$

The thermodynamic driving force of the development of the ensemble of
clusters along the valley of the Gibbs free energy is the decrease of
Δ G due to the decrease of the number of clusters. Going over to the
idealized ensemble of s identical clusters with the radius R (corresponding to the mean value of the radius in the real ensemble)
$\partial\Delta G/\partial r_\alpha^{(j)}$ has to be replaced, therefore, by

$$s\,\frac{\partial\Delta G}{\partial r_\alpha^{(j)}} \longrightarrow \frac{d\Delta G}{ds}\,\frac{ds}{dR} \tag{6.38}$$

where Δ G is determined now instead of eq. (6.23) by eq. (6.24). With
eqs. (6.28) – (6.31) the following differential equation for the mean
radius of the ensemble of clusters is obtained:

$$\frac{dR}{dt} = \omega\,\frac{Dc_\beta\sigma}{c_\alpha^2 kT}\,\frac{1}{R1}\,(1 + Z^{-1}) \tag{6.39}$$

The numerical factor ω , introduced in addition in eq. (6.39), will
be determined later.
The variation of the number of clusters with time can be calculated
by the generalized Gibbs-Thomson equation (6.27) or deriving this

equation with respect to time, by the differential equation (6.40)

$$\frac{d}{dt}\left\{\ln\left(\frac{sV_\alpha}{V}\right)\right\} = -\frac{1}{Z}\frac{d}{dt}\left\{\ln(R^3)\right\} \quad,\quad V_\alpha = \frac{4\pi}{3}R^3 \quad. \tag{6.40}$$

The total mass concentrated in the clusters or the total volume of the new phase and, therefore, the critical radius are monotonicly increasing functions of time (see eqs. (6.35) and (6.40)).
In the vicinity of the highest point of the valley of the thermodynamic potential, given by eq. (6.32), the quantity Z equals minus one and, ·consequently, the mean radius of the ensemble of clusters increases slowly. Such a relatively slow increase of the mean radius of the ensemble of clusters in the first stage of the ripening process is, indeed, observed in computer simulations (e.g. KOCH, DESAI, ABRAHAM, 1983; GUNTON, SAN MIGUEL, SAHNI, 1983). This period of slow growth is followed by a period of rapid increase of the total volume of the new phase, leading to a rapid increase of the absolute value of the quantity Z (see also: SCHMELZER, 1985d). Afterwards the total volume of the new phase remains nearly constant ($Z^{-1} \simeq 0$, $c_\beta \simeq c'$) and the time development of the mean radius of the ensemble of clusters is described
by
$$\frac{dR}{dt} = \omega\,\frac{D\sigma c'}{c_\alpha^2 kT}\,\frac{1}{R1} \quad. \tag{6.41}$$

Neglecting in this asymptotic region surface effects and the term sV_α/V compared with one, from the Gibbs-Thomson equation (6.27) we obtain, approximately,

$$\ln\frac{c}{c'} = s\,\frac{n_\alpha}{n_2} \tag{6.42}$$

leading with eq. (6.41) to the following expression for s as a function of time:

$$s = \frac{n_2}{c_\alpha V_\alpha}\,\ln\frac{c}{c'} \quad. \tag{6.43}$$

For diffusion limited growth the time evolution of the mean radius of the ensemble of clusters and its number ·is given in the asymptotic region, therefore, by

$$R^3 = \omega\,\frac{3c'D\sigma}{c_\alpha^2 kT}\,t \quad;\quad s = \frac{n_2 c_\alpha}{c'}\,\frac{kT\,\ln\frac{c}{c'}}{4\pi D\sigma\omega}\,\frac{1}{t} \tag{6.44}$$

and, in agreement with WAGNER (1961), we get for interface kinetic limited growth

$$R^2 = 2 \frac{c'}{c_\alpha^2} \frac{D\sigma\omega}{kTl_0} t; \quad s = \frac{3n_2 c_\alpha^2}{4\pi}(\ln \frac{c}{c'})(\frac{kTl_0}{2c'D\sigma\omega})^{3/2} \frac{1}{t^{3/2}} . \quad (6.45)$$

A comparison of the eqs. (6.44) with the results of Lifshitz and
Slyozov (eqs. (6.18), (6.29)) shows a qualitative agreement. This pro-
ves first, that the evolution of the ensemble of clusters along the
valley of the characteristic thermodynamic potential corresponds, indeed, to the process of Ostwald ripening. So, the eq .(6.32) gives an
estimation of the lowest possible value of the mean radius of the ensemble of clusters and the value of the number of clusters for which
the process of Ostwald ripening may start. Moreover, since the eqs.
(6.39) and (6.40) are valid for the whole ripening process, the method
outlined here allows the theoretical description of the whole ripening
process, including the initial stage.
In the asymptotic region our approach is also in quantitative agreement with the results of Lifshitz, Slyozov and Wagner if the introduced
in eq. (6.39) numerical factor ω is set equal to 8/27. The quantitative deviation (for $\omega = 1$) is due first to the fact, that in the growth
equation (6.11) used by Lifshitz and Slyozov, c_β is replaced by c',
and second, to the assumption of an idealized ensemble of s identical
clusters underlying our method. Since in the asymptotic region the results of the Lifshitz-Slyozov theory are exact the correction factor ω
will be determined as

$$\omega = 8/27 \qquad\qquad (6.46) .$$

to get a coincidence of both approaches in this stage of the ripening
process.
Numerical integrations of the eqs. (6.39) and (6.40) lead to results
as presented in Fig. 6.1. In the first period of the process of Ostwald ripening both the total volume $V_{\alpha t}$ and the total surface area of
the ensemble of clusters increase. So, in this period the thermodynamic driving force of the ripening process is the decrease of the Gibbs
free energy due to the increase of the total mass of the new phase.

In the second stage, described by the asymptotic solutions the total volume of the new phase remains nearly constant while the total
surface area decreases as

$$A_t = n_2 \ln \frac{c}{c'} \left[\frac{9kT}{\omega c_\alpha c' D\sigma} \right]^{1/3} \frac{1}{t^{1/3}} \qquad\qquad (6.47)$$

for diffusion limited growth and as

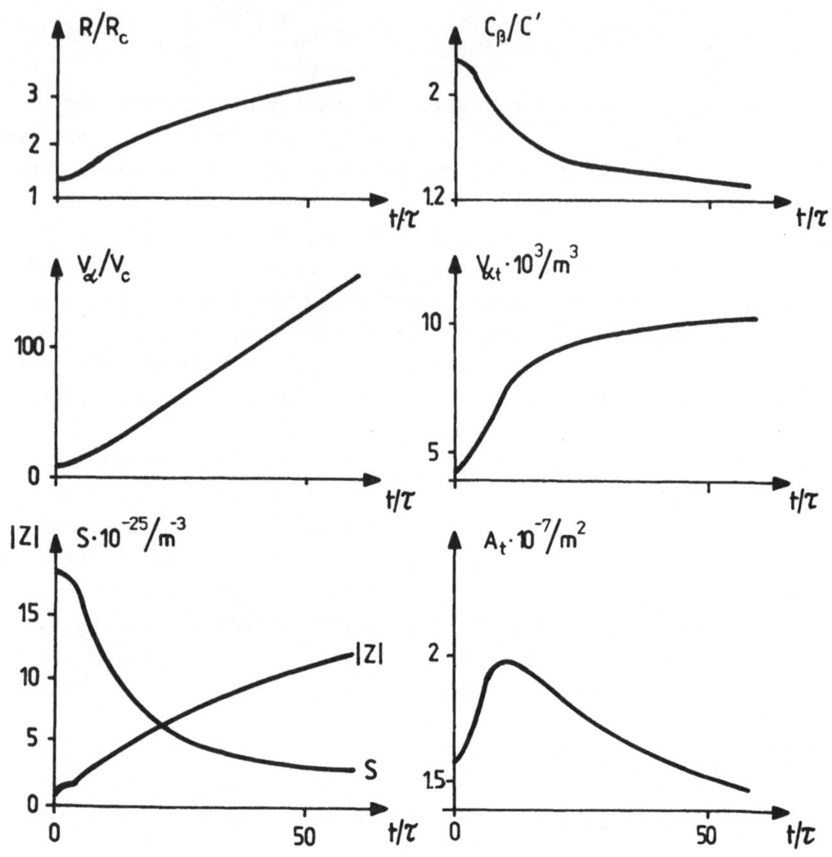

Fig. 6.1:
Time development of the mean radius R/R_c, the supersaturation c_β/c', the mean volume of the clusters V_α/V_c, the total volume of the new phase $V_{\alpha t}$, the total surface area A_t and the number s of the clusters. R_c is here the critical radius for one cluster in an infinite system, V_c its volume. The values of the parameters are given by $c = 4.2 \cdot 10^{26} m^{-3}$, $c_\alpha = 2.3 \cdot 10^{28} m^{-3}$, $c' = 1.4 \cdot 10^{26} m^{-3}$, $\sigma = 0.08\ Nm^{-1}$, $D = 1.2 \cdot 10^{-5} m^2 s^{-1}$, $T = 730\ K$, which correspond to the experimental case of Ostwald ripening of AgCl clusters in a sodium borate melt (PASCOVA, GUTZOW, 1983). The radius of the critical cluster is equal to $R_c = 6.28 \cdot 10^{-10} m$, τ is determined by $\tau = c_\alpha^2 kTR_c^2 \omega/(2\sigma cD)$. It is equal to $\tau = 0.48 \cdot 10^{-2} s$ for the assumed values of the parameters.

(From: SCHMELZER, 1985a)

156

$$A_t = 3n_2 \ln \frac{c}{c'} \left[\frac{kTl_0}{2c'D\sigma\omega} \right]^{1/2} \frac{1}{t^{1/2}} \tag{6.48}$$

for interface kinetic limited growth. In this asymptotic region, the thermodynamic driving force of the process is the decrease of the surface contributions to the Gibbs free energy.

Based on eq. (6.20) by the same method the process of growth of in-compressible drops with a molar density ϱ_α and the process of Ostwald ripening of an ensemble of drops in a binary gaseous mixture under a constant pressure and constant temperature has been described (SCHMELZER, 1985a; SCHMELZER, SCHWEITZER, 1987b). Assuming, again, that the drops consist mainly of the second component and denoting its chemical potential in the medium by $\mu_{2\beta}$, its molar density by $\varrho_{2\beta}$, the growth equation for one drop is obtained in the form

$$\frac{dr_\alpha}{dt} = \frac{2D\sigma\varrho_{2\beta}}{\varrho_\alpha^2 R_g T} \frac{1}{r_\alpha} \left[\frac{1}{r_c} - \frac{1}{r_\alpha} \right] \quad r_c = \frac{2\sigma}{\varrho_\alpha(\mu_{2\beta}-\mu')-(p-p')} \quad . \tag{6.49}$$

The time evolution of the ensemble of drops is described by

$$\frac{dR}{dt} = \omega \frac{\varrho_{2\beta}}{\varrho_\alpha^2} \frac{D}{R_g T} \frac{\sigma}{r_\alpha^2} (1 + Z^{-1}) \tag{6.50}$$

$$\frac{d}{dt} \left[\ln(sn_\alpha) \right] = - \frac{1}{Z} \frac{d}{dt} \left[\ln(R^3) \right] ; \quad Z = - \frac{3R\varrho_\alpha sV_\alpha}{2\sigma} \frac{n_1}{(n_1+n_2-sn_\alpha)^2} \frac{\partial\mu_{2\beta}}{\partial x_\beta} \quad .$$

Based on the results of the thermodynamic investigation of the process of condensation in a one-component closed isochoric system the growth equations for one drop and the ensemble of drops in the stage of Ost-wald ripening are given by eqs. (6.49) and (6.50), again, where $\mu_{2\beta}$ has to be replaced by μ_β . Z is given in this case by eq. (3.72). (For the details and the discussion of special cases see: SCHMELZER, 1985 a,e; SCHMELZER and ULBRICHT, 1986).

6.5. Ostwald Ripening and the Relations to the Theory of Self-organization

The evolution of chemical and biological species can be described in a deterministic sense by a system of nonlinear ordinary differential equations (dynamic system) of the form

$$\frac{dx_i}{dt} = f_i(x_1, x_2, \ldots, x_n, u) \qquad \text{for } i = 1, 2, \ldots, m \tag{6.51}$$

x_i are the concentrations of the different species, raw materials, etc.; u is a set of external parameters (control parameters).
In the special case, that there exists no direct interaction between the species, the competition is due mainly to the limited amount of raw material, the constant overall particle number or other limiting the growth factors. Then it can be shown, that widely independent of the type of the growth function f_i (eq. (6.51)) the number of different species, which can survive the process of competition, cannot be greater than the number of limiting growth factor (SCHMELZER, 1979; EBELING, SCHMELZER, 1980a,b; SCHMELZER, EBELING, 1980; SCHMELZER, EBELING, FEISTEL, 1981). In biology, such a statement is known as the principle of Gauze-Volterra.
Examples of equations (6.51) describing parallel reactions are (SCHMELZER, 1979; EBELING, FEISTEL, 1982)

$$\dot{x}_i = u_i \, A \, x_i - u_i' \, x_i \qquad i = 1, 2, \ldots, m$$

$$\dot{A} = \phi_A - \sum_{l=1}^{m} u_i \, A \, x_i \tag{6.52}$$

which represent competition and selection due to the limited amount of raw material A. ϕ_A is the constant input flux of raw material, u_i and u_i' are constant parameters which are related to the properties of the species. Defining

$$E_i = u_i \, A - u_i' \tag{6.53}$$

as replication rate the stability analysis shows that the species with $u_s'/u_s < u_i'/u_i$ survive in the system. Therefore the stationary solution is given by

$$E_s = 0 \quad ; \quad x_s \neq 0$$

$$E_i \neq 0 \quad ; \quad x_i = 0 \qquad \text{for all } i \neq s \tag{6.54}$$

Another example of a competitive system is the so-called Fisher-Eigen model of prebiotical evolution (EIGEN, 1971) written as

$$\dot{x}_i = (E_i - <E>) \, x_i \qquad i=1,2,\ldots,m \qquad\qquad (6.55)$$

with

$$<E> = \sum_{i=1}^{m} E_i x_i / \sum_{i=1}^{m} x_i \qquad\qquad (6.56)$$

The time dependent quantity $<E>$ (eq. (6.56)) is called the mean reproduction rate or average population fitness. Summing eqs. (6.55) the
constraint is given by the constant overall organization

$$\sum_{i=1}^{m} x_i = C = \text{const} \qquad\qquad (6.57)$$

The system of coupled equations (6.55) has been investigated by many
authors (see, e.g., EIGEN and SCHUSTER, 1979; MAHNKE, 1979; EBELING
et al., 1984). The time dependent solutions $x_i(t)$ can be calculated
(Fig. 6.2).
The stationary state

$$x_s = C \qquad \text{with } E_s = \text{Max}\{E_1,E_2,\ldots,E_m\}$$
$$x_i = 0 \qquad \text{for all } i = s \qquad\qquad (6.58)$$

shows, that the species with the highest reproduction rate (selection
value) $E_s > E_i$ will increase to the finite concentration C (eq. (6.57))
and all others must die out. Eq. (6.55) describes explicitely a selection procedure: Under stated selection constraints (eq. (6.57)) the
population numbers of all but one species will diminish.

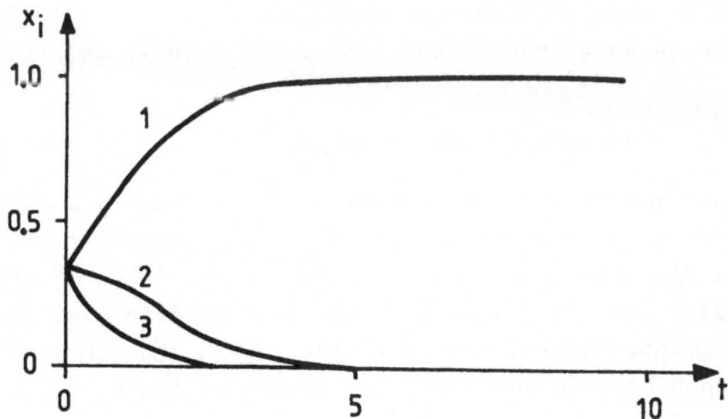

Fig. 6.2:
Numerical integration of the Fisher-Eigen model (eq.(6.55)) for three
sorts. Replication rates (eq. (6.53)): $E_1 = 3$; $E_2 = 2$; $E_3 = 1$
(From: MAHNKE, 1979)

In an analogous way, Ostwald ripening represents a competitive growth
of the different clusters of the ensemble due to the limited amount of
particles (monomers) the clusters are formed of. In the late stage of
phase transition when the supersaturation is small and near to its
equilibrium value, we observe that larger clusters grow at the expense
of smaller ones which disappear. In the final stable equilibrium state
only one drop will remain in the system. As was pointed out by MAHNKE
and FEISTEL (1985) (see also: ULBRICHT et al., 1986; MAHNKE, 1986,
1987) the process of Ostwald ripening or coarsening is closely related
to the competitive evolution of selfreplicative units in the Eigen-
Fisher model (eq. (6.55)) with the global closure condition (eq.
(6.57)).

Now, the dynamics of Ostwald ripening is investigated in terms of kinetic equations recently derived for the intermediate as well as the
late stage of first-order phase transitions. Starting with the basic
equations of the competitive growth problem (Chapter 6.1.) we derive
a set of ordinary nonlinear differential equations governing the constrained growth of a droplet ensemble with m components. Therefore,
the continuous cluster distribution density (eq. (6.1)) is written in
the discrete case for $m \geqslant 1$ different kinds of droplets as

$$f(n,t) = \frac{1}{V} \sum_{i=1}^{m} N_i(t) \ \delta(n - n_i(t)) \qquad (6.59)$$

where $N_i(t)$ is the number of clusters and $n_i(t)$ the size (number of
bounded monomers) of the clusters of class i at time t. Inserting the
ansatz of the cluster size distribution function (eq. (6.59)) into the
continuity equation (6.8) without production-decay term we get for
$i = 1, 2, \ldots, m$

$$N_i(t) = N_i(0) \quad ; \quad dn_i/dt = v_i(n_1, \ldots, n_m, c_\beta) \qquad (6.60)$$

Eq. (6.60) describes the deterministic motion of the droplet ensemble
in the m-dimensional size space n_1, $n_2, \ldots, n_i, \ldots, n_m$. Since in eq.
(6.8) there are no diffusion contributions like in the Fokker-Planck
equation (see Chapter 4.5.) the height of the distribution over the
deterministic trajectory remains constant, that means the number of
clusters is unchanged (eq. (6.60)). For simplicity we fix $N_i = N_i(0) = 1$
at the initial time $t = 0$. In this approach the index i plays the role
of an integer for labeling the different kinds of drops. This label
do not change when the cluster is evolving.

Now we calculate the dynamic system (eq. (6.60)) explicitly using
the growth rate (eq. (6.6)) in the interface kinetic limited case given by

$$\frac{dn_i}{dt} = v(n,c_\beta) = D \; l_o^{-1} \; A(n)(c_\beta(t) - c_{eq}(n)) \qquad (6.61)$$

with

$$c_{eq}(n) = c' \exp(l_o \, k(n)) \approx c' \, (1 + l_o \, k(n)) \qquad (6.62)$$

where

$$l_o \;\; = 2\bar{G}\,(c_d\,k_B T)^{-1} \approx 1 \text{ nm} \qquad \text{capillary length (eq.(4.95))} \quad (6.63)$$

$$A(n) = 4\pi\,(c_d\,4\pi/3)^{-2/3}n^{2/3} \qquad \text{surface of a size-n-droplet} \quad (6.64)$$

$$k(n) = (c_d\,4\pi/3)^{1/3}n^{-1/3} \qquad \text{curvature of a size-n-} \quad (6.65)$$
$$\text{droplet}$$

The growth law (eq. (6.61)) embodies much of the physics of the process of condensation or evaporation of monomers changing the cluster size. It takes into consideration, that the equilibrium monomer concentration $c_{eq}(n)$ (eq. (6.62)) over a curved surface with curvature (eq. (6.65)) $k(n) = 1/r_n$, is higher in comparison with the equilibrium value c' over a planar surface. This curvature dependent expression (Gibbs-Thomson equation (6.62)) indicates that atoms will flow from regions of high to low curvature.

Using the supersaturation y (compare eq. (1.1) for y_o) in the approximation

$$\ln y(t) = \ln p/p' = \ln c_\beta/c' \approx (c\,(t) - c')/c' \qquad (6.66)$$

eq. (6.61) can be written in the equivalent form

$$\dot{n} = D \; c' \; l_o^{-1} \; A(n)(\ln y(t) - l_o \, k(n)) \qquad (6.67)$$

and after introducing the critical droplet size n_{cr} by

$$n_{cr} = (c_d\,4\pi/3)(l_o/\ln y(t))^3 \qquad (6.68)$$

we get

$$\dot{n} = D \; c' \; A(n)(k(n_{cr}) - k(n)) \qquad (6.69)$$

which represents the well-known droplet kinetics already given in Chapter 4 (compare eq. (4.91)). Because of the depletion of the monomers in the finite system the critical size n_{cr} is determined by the whole droplet ensemble and changing in time.

Now we want to show that the curvature of the critical cluster $k(n_{cr})$ in eq. (6.69) is given by the mean curvature $<k>$ which relates eq. (6.69) to the Eigen-Fisher equation (6.55). Inserting the growth rates (eq. (6.61)) for all clusters $i=1,2,\ldots,m$ into the closure condition of the finite system (conservation of monomers, eq. (6.7)) after taking the time derivative we get

$$\dot{c}_\beta = - \sum_{i=1}^{m} (N_i/V)\dot{n}_i = - D\, l_0^{-1} \sum_i (N_i/V)A(n_i)(c_\beta - c' - c'l_0 k(n_i))$$

$$= - D\, l_0^{-1} A_\alpha/V\,(c_\beta - c') + D\, c' \sum_i N_i A(n_i)k(n_i)/V \qquad (6.70)$$

Substituting $(c_\beta - c')$ from eq. (6.70) into eq. (6.61) we obtain the
final set of ordinary nonlinear differential equations governing the
constrained growth or Ostwald ripening regime of an ensemble of m
droplets $(n_1,\ldots,n_m)$ and monomers $(c_\beta$, eq. (6.2))

$$\dot{n}_i = D\, l_0^{-1} A(n_i) \left[c'l_0(\langle k \rangle - k(n_i)) - \frac{Vl_0}{D}\frac{1}{A_\alpha}\,\dot{c}_\beta \right] \qquad (6.71)$$

$$i = 1,2,\ldots,m$$

with

$$\dot{c}_\beta = - D\, l_0^{-1} A\, /V \left[c_0 - c_\alpha V_\alpha/V - c' - l_0 c'\langle k \rangle \right] \qquad (6.72)$$

where

$$A_\alpha(t) = \sum_i N_i A(n_i) \qquad \text{total surface of all droplets} \qquad (6.73)$$

$$V_\alpha(t) = c^{-1} \sum_i N_i n_i \qquad \text{total volume of all droplets} \qquad (6.74)$$

$$\langle k \rangle(t) = \frac{\sum_i k(n_i)N_i A(n_i)}{\sum_i N_i A(n_i)} \qquad \text{mean curvature} \qquad (6.75)$$

and $c_0 = N_m/V$ = const being the total number of monomers (eq. (4.4))
per volume.

The kinetic equations (6.71, 6.72) (MAHNKE, FEISTEL, 1985) describe
the rapid growth of the droplet ensemble (second term $\dot{c}_\beta$ on r.h.s. of
eq. (6.71)) and the slow selection process between the droplets
(first term on r.h.s. of eq. (6.71)). Scaling with $z = n/n_{cr}$ the developed equations they can be transformed in a suitable dimensionless
form given elsewhere (MAHNKE, 1986; ULBRICHT et al., 1986).
Numerical solutions of the coupled nonlinear equations (6.71, 6.72)
are given in the following Figures. In Fig. 6.3a the time evolution
of a droplet distribution with three classes of drops (m=3) is shown.
The time is scaled by the unit $t_0 = c_\alpha l_0^2/(c'D) = 1.16$ ns. In a short
time regime the size of all drops is increasing and the two phase system will reach a so-called internal equilibrium $(\dot{c}_\beta \approx 0)$, a nonstationary situation where the liquid (α-phase) is in a quasi-equilibrium with its surrounding vapour (β-phase). Since the raw material
(monomers) is limited at the end of this growth stage the total volume
V_α (eq. (6.74)) (or mass) of the new phase is very close but not equal
to its equilibrium value (Fig. 6.5c). In the final long-time regime
a competitive ripening process takes place which minimizes the total
surface A_α (eq. (6.73)). At the end of the coarsening process the dy-

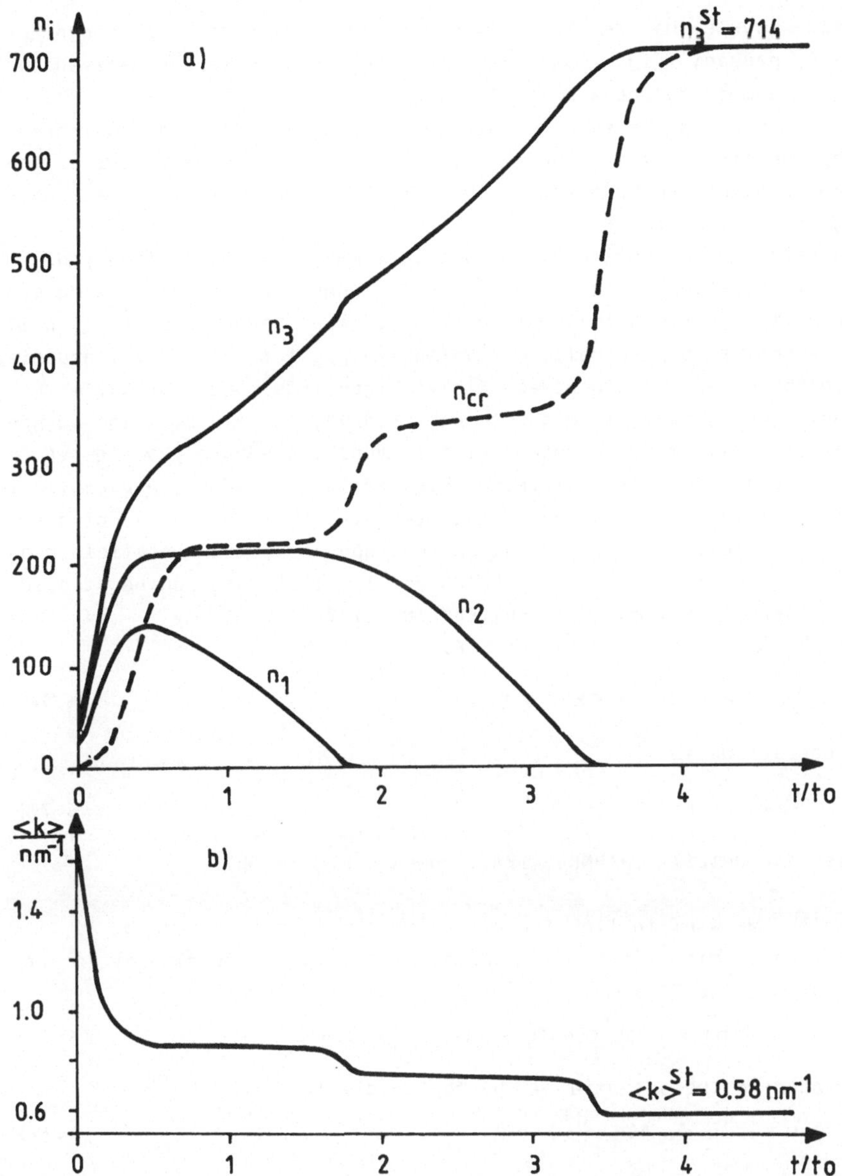

Fig. 6.3: Growth and ripening of a droplet ensemble with three size classes. Vapour: water, T = 293 K, y_o = 5.7

a) Time evolution of the droplets with sizes n_1, n_2, n_3 (initial values: $n_1(0)=40$, $n_2(0)=50$, $n_3(0)=75$) in comparison with the critical value n_{cr}
b) Time dependence of the mean curvature of the evolving droplet ensemble (eq. (6.75))
(From: MAHNKE, 1987)

163

namic system has reached a stable fixed point where only one droplet is present (Fig. 6.3a). The other smaller clusters dissolved to
give monomers to the winner, the biggest drop.
Fig. 6.3b shows the mean curvature $<k>$ (eq. (6.75)) as a non-increa-
sing function of time (MAHNKE, 1987). The step-like behaviour of the
mean curvature corresponds to the shrinkage of small droplets, compare
Fig. 6.3a with Fig. 6.3b.
Focussing the attention on the simularities between the theory of
Ostwald ripening presented here and the theory of selforganizing systems it is obvious that our derived dynamic   equations (6.71, 6.72)
are a realistic physically motivated example for natural selforganiza-
tion.Note that in both cases (Fisher-Eigen model (eq. (6.55)) with
population average fitness $<E>$ (eq. (6.56)), Ostwald ripening of precipitates with mean curvature of the droplet ensemble $<k>$ (eq. (6.75))
the interaction of the different species is modeled by an overall di-
lution flux due to the constraints (eq. (6.75) , eq. (6.7)) of the
finite system. This consideration corresponds to the mean-field concept of many-body physics (EBELING et al., 1984) and can be studied
in different systems (FEISTEL, EBELING, 1976).
For the mean productivity (eq. (6.56)) we obtain

$$\frac{d}{dt} <E> = <E^2> - <E>^2 \geqslant 0 \tag{6.76}$$

in comparison to

$$\frac{d}{dt} <k> \leqslant 0 \tag{6.77}$$

(see, for details, MAHNKE, 1987; compare Fig. 6.3b).

Finally, we want to link the given results to the statements of the
previous Chapters. First we consider the case of one cluster in the
vapour (eq. (6.71)):

$$\dot{n} = f(n) = D \ l_0^{-1} A(n)(c_0 - n/V - c' - l_0 c' k(n)) \ . \tag{6.78}$$

Introducing the potential Φ by the gradient

$$\dot{n} = - \frac{D}{l_0} \frac{c'}{kT} A(n) \frac{d\Phi}{dn} \tag{6.79}$$

as it was done in Chapter 5.3.2. (compare eq. (5.20)), we obtain

$$\Phi(n) = - kT \ (\ln y_0) \ n + (kT/(2Vc')) \ n^2 + \mathfrak{G} A(n) \tag{6.80}$$

consisting of a nonlinear volume part and the usual surface contribution. Fig. 6.4 shows the bistable potential and the fixed points
(MAHNKE, 1986), see also Fig. 3.1.

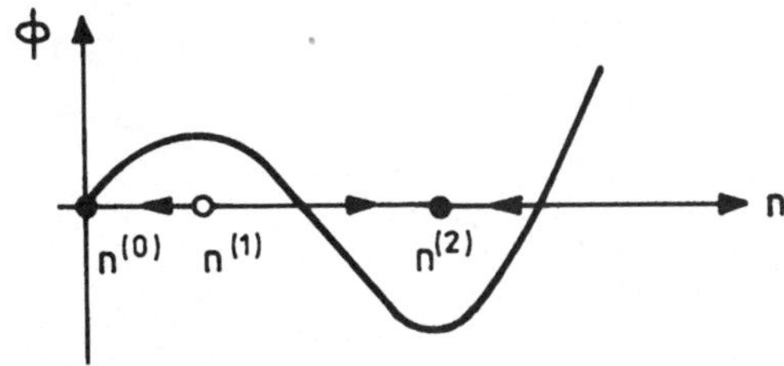

Fig. 6.4:
Potential Φ (eq. (6.80)) versus cluster size n. The fixed points correspond to the vapour phase ($n^{(0)}$ = 0, stable), the critical droplet size ($n^{(1)} = n_{cr}$, unstable) and the liquid drop ($n^{(2)} = n_{st}$, stable)

Now we investigate the two-dimensional case of the dynamic system (eq. (6.71), i=1,2), a situation where two clusters of sizes n_1 and n_2 are in a bath of monomers. Fig. 6.5 shows the time dependence of overcritical clusters, the monomer concentration c_β or supersaturation y and the overall quantities A_α (eq. (6.73)) and V_α (eq. (6.74)) by numerical integration of eqs. (6.71,6.72) for a system containing N_m = 300 particles. As already discussed (Fig. 6.3) we distinguish constrained growth (short-time regime) and sharp selection (long-time regime).

Nowadays, fixed point analysis is a routine technique for studying the long-term behaviour of dynamic systems. It can be found in a number of textbooks, see, e.g., EBELING and FEISTEL (1982), EIGEN and SCHUSTER (1979). As usually done by fixed point analysis the behaviour of the dynamic system becomes quite clear investigating the topology of the state space with all fixed points. According to the response of the system to small changes in the variables the stability of the stationary states is investigated. We can distinguish different classes of fixed points: stable points or sinks (attractors, ●), saddle points (⊕) and unstable points or sources (O) representing the locally highest points. After long enough time every realistic dynamic system will approach an attractor. Within a given basin the approach to the attractor is independent of the particular initial conditions.
In our example the state space of the cluster sizes is a simplex $\{n_1,n_2 \mid 0 \le n_1+n_2 \le N_m\}$ with a number of stationary states of different stability and an interesting topologic feature (Fig. 6.6). The trajectories indicate the flow from unstable to stable states. The attractor at zero ($n_1^{(0)} = n_2^{(0)} = 0$) corresponds to the pure vapour phase, its basin is bounded by the separatrices (dotted lines) between the saddles on the boundaries and the source on the principal diagonal.

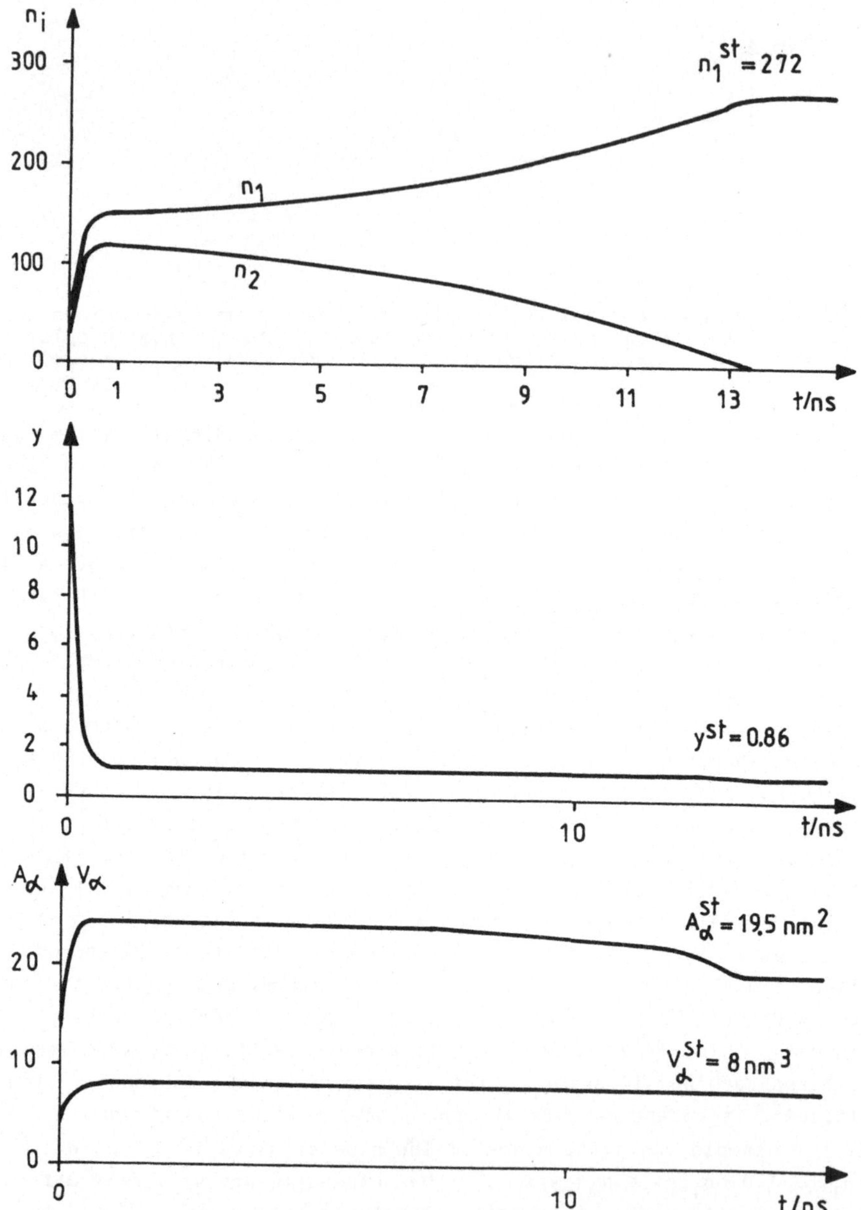

Fig. 6.5: Time dependence of droplet sizes n_1, n_2 (a), supersaturation y (b), total droplet surface A_α and total droplet volume V_α (c) in the growth and ripening stage.
Vapour: water, $T = 293$ K, $N_m = 300$ particles, $V = 26000$ nm^3

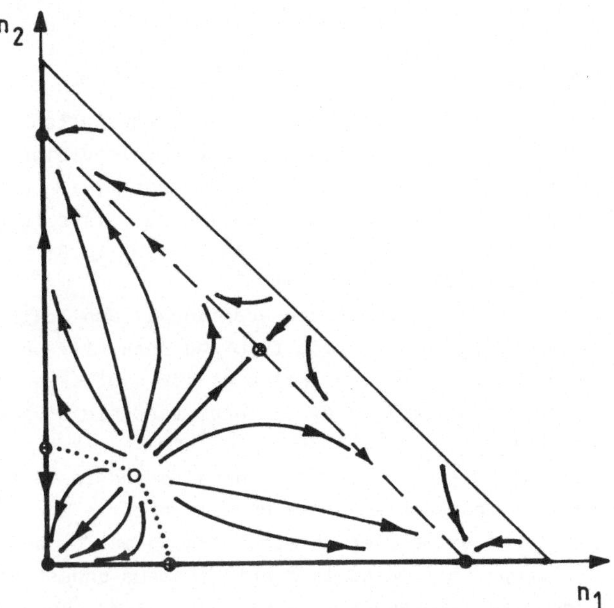

Fig. 6.6:
Schematic plot of the flow in the two-dimensional space of cluster sizes (called simplex) with unstable ($\bigcirc$), stable ($\bullet$) and saddle ($\oplus$) fixed points.

The separatrices represent the critical cluster sizes, in particular the unstable point ($n_1^{(1)} = n_{cr}$, $n_2^{(1)} = 0$) is known from the one-dimensional case (Fig. 6.4), but it changes its type from node to saddle. On the principal diagonal ($n_1 = n_2$, identical clusters) there is not only the unstable node which is related to the highest point of the thermodynamic potential (Fig. 3.18) but also a further saddle. The line between this saddle and the stable equilibrium situations ($n_1^{(2)} = n_{st}$, $n_2^{(2)} = 0$ or $n_1^{(2)} = 0$, $n_2^{(2)} = n_{st}$) corresponds to minima of the thermodynamic potential (dashed curves in Fig. 6.6, compare the valley of the free energy in Fig. 3.18). Overcritical clusters evolves in such a way, first they reach the valley of the thermodynamic potential (growth regime) and then in a long-time regime of selection they will approach the stable fixed point going downstairs in the valley (MAHNKE, FEISTEL, 1985). After about one nanosecund the system has reached internal equilibrium, but it takes 10 or 100 times more for the ripening process depending on the diffusion constant D.

7. Growth of Bubbles in Finite Systems

7.1. The Model

Though the main theoretical concepts for the description of the formation and growth of the bubbles are the same as for clusters with a higher compared with the surrounding medium density the case of formation and growth of bubbles is investigated much less (see, e.g., WARD et al., 1982, 1984; KAGAN, 1960; RAO et al., 1978; SCHMELZER, SCHWEITZER, 1987a; BLANDER et al., 1971).

At part this situation is due to the additional complications in the theoretical description which arise from the necessity to use two independent variables for the description of the bubble instead of one for drops, since the assumption of an incompressible cluster phase is not applicable yet.

Here we would like to show, that the general scenario of the phase transition remains the same also for bubbles despite these additional complications. As an example we consider the formation and growth of bubbles in one-component closed isochoric systems under isothermal conditions (see also, SCHMELZER, 1986b). Under these conditions the formation of the bubbles is possible only if the liquid has a lower compared with the equilibrium value ϱ' initial molar density $\varrho(\varrho < \varrho')$ The density ϱ' corresponds to the equilibrium pressure p' for the coexistence of both phases at a planar interface.

If the pressure p is decreased the volume of the liquid increases until p becomes equal to zero. A further increase of the volume of the liquid is possible only if the system is extended by external forces. This is equivalent to negative values of the pressure p.

The possibility of negative values of the pressure p do not alter the general thermodynamic relationships and results of the thermodynamic analysis (SCHMELZER, 1986b). In particular, the correction term to the work of formation of the critical bubbles is given by

$$\Delta W = - \frac{V}{2} \frac{\partial \mu}{\partial \varrho} (\Delta \varrho)^2 \tag{7.1}$$

It is negative (or equal to zero for $\Delta \varrho = 0$).

The change of the parameters of the critical or stable bubbles are also expressed by eqs. (3.60), (3.61). With an increasing size of the bubbles its density increases.

7.2. Thermodynamic Analysis

Assuming s bubbles are formed in the system and neglecting a possible
curvature dependence of the surface tension the change of the free
energy of the system due to the formation of the bubbles is given by

$$\Delta F = \sum_{j=1}^{s} ((p_\beta - p_\alpha^{(j)}) V_\alpha^{(j)} + (\mu_\alpha^{(j)} - \mu_\beta) n_\alpha^{(j)} + \sigma A^{(j)}) +$$

$$+ (p - p_\beta) \, V \, + \, (\mu_\beta - \mu) n \tag{7.2}$$

The variation of the free energy of the heterogeneous system in reversible processes reads

$$dF_{het} = \sum_{j=1}^{s} ((p_\beta - p_\alpha^{(j)} + \sigma \frac{dA^{(j)}}{dV_\alpha^{(j)}}) \, dV_\alpha^{(j)} + (\mu_\alpha^{(j)} - \mu_\beta) dn_\alpha^{(j)}) \tag{7.3}$$

In the special case only one bubble is present in the system these
equations are reduced to

$$\Delta F = (p_\beta - p_\alpha) V_\alpha + \sigma A + (\mu_\alpha - \mu_\beta) n_\alpha + (p - p_\beta) V + (\mu_\beta - \mu) \, n \tag{7.4}$$

$$dF_{het} = (p_\beta - p_\alpha + \sigma \frac{dA}{dV_\alpha}) \, dV_\alpha + (\mu_\alpha - \mu_\beta) \, dn_\alpha \tag{7.5}$$

From eq. (7.3) it follows immediately that the necessary equilibrium
conditions are given, again, by

$$p_\alpha^{(j)} - p_\beta = 2\sigma / r_\alpha^{(j)} \tag{7.6}$$

$$\mu_\alpha^{(j)} = \mu_\beta \tag{7.7}$$

Taking into account the Gibbs-Duhem relations a derivation of ΔF with
respect to the independent variables $n_\alpha^{(j)}$ and $V_\alpha^{(j)}$ leads to the same
results.
The necessary equilibrium conditions can be fulfilled only if all
bubbles are characterized by the same parameters n_α and V_α. In this
case eq. (7.2) is reduced to

$$\Delta F = -s(p_\alpha - p_\beta) V_\alpha + (\mu_\alpha - \mu_\beta) s n_\alpha + s \sigma A + (p - p_\beta) V + (\mu_\beta - \mu) \, n \tag{7.8}$$

ΔF depends on the variables n_α, V_α and s, now. A derivation of eq.
(7.8) with respect to these independent variables yields

$$\left(\frac{\partial \Delta F}{\partial n_\alpha}\right)_{V_\alpha,s} = -s\,V_\alpha\left[\frac{\partial p_\alpha}{\partial n_\alpha} - \varrho_\alpha\frac{\partial \mu_\alpha}{\partial n_\alpha}\right] + V_\beta\left[-\frac{\partial p_\beta}{\partial n_\alpha} + \varrho_\beta\frac{\partial \mu_\beta}{\partial n_\alpha}\right] + s\,(\mu_\alpha - \mu_\beta) \tag{7.9}$$

$$\left(\frac{\partial \Delta F}{\partial V_\alpha}\right)_{n_\alpha,s} = -s\left[p_\alpha - p_\beta - \sigma\frac{\partial A}{\partial V_\alpha}\right] - s\,V_\alpha\left[\frac{\partial p_\alpha}{\partial V_\alpha} - \varrho_\alpha\frac{\partial \mu_\alpha}{\partial V_\alpha}\right] + V_\beta\left[-\frac{\partial p_\beta}{\partial V_\alpha} - \varrho_\beta\frac{\partial \mu_\beta}{\partial V_\alpha}\right] \tag{7.10}$$

$$\frac{d\,\Delta F}{ds} = \left(\frac{\partial \Delta F}{\partial V_\alpha}\right)_{n_\alpha,s}\frac{d V_\alpha}{ds} + \left(\frac{\partial \Delta F}{\partial n_\alpha}\right)_{V_\alpha,s}\frac{d n_\alpha}{ds} + \left(\frac{\partial \Delta F}{\partial s}\right)_{n_\alpha,V_\alpha} \tag{7.11}$$

Eq. (7.11) is valid for the extrema of ΔF, since the equilibrium con-. ditions determine n_α and V_α as a function of s. Moreover, for the extrema

$$\left(\frac{\partial \Delta F}{\partial n_\alpha}\right)_{V_\alpha,s} = \left(\frac{\partial \Delta F}{\partial V_\alpha}\right)_{n_\alpha,s} = 0 \tag{7.12}$$

eq. (7.11) results in

$$\frac{d\,\Delta F}{ds} = \frac{\partial \Delta F}{\partial s} = \frac{1}{3}\sigma A \tag{7.13}$$

For an ensemble of s identical bubbles the increase of their number s is equivalent to a decrease of the volume per bubble. In agreement with eqs. (3.60), (3.61) an increase of the number of bubbles results therefore in an increase of the radius of the critical bubbles and a decrease of the size of the stable (for fixed values of s) bubbles. Consequently, we obtain a behaviour, again, as shown for the case of drops in Figs. 3.15, 3.17. The general scenario of the phase transition remains, therefore, the same.

7.3. Kinetic Descritpion of Nucleation and Growth of the Bubbles

If the steady-state nucleation rate is known depletion effects can be taken into account by the quasi-steady-state approximation discussed in chapter 5.2. So we can restrict ourselves here to the kinetic description of the growth of the already formed bubbles.
According to eq. (5.20) the growth of a cluster of the new phase with the radius r can be written as

$$\frac{d n_\alpha}{dt} = -\frac{4\pi r_\alpha^2\,D\,\varrho_\beta}{R_g T}\frac{1}{l}\frac{\partial \Phi}{\partial n_\alpha} \tag{7.14}$$

ρ_β is the molar density of the segregating particles in the medium surrounding the cluster at a sufficiently large distance from it, R_g the universal gas constant.

This equation was derived based on the assumption, that one independent variable is sufficient for the description of the state of the cluster. As it was already mentioned, this is not the case for bubbles, since according to eq. (7.5) two independent parameters are needed.

But since the mechanical equilibrium is established much more rapidly compared with the diffusion equilibrium we can assume that at each moment eq. (7.6) is fulfilled. This equation gives an additional relation between n_α and V_α and, consequently, V_α can be considered as the only independent variable, now.

Taking into account this additional condition eq. (7.14) is applicable, again, and the kinetic equation for the description of the evolution of the radius of one bubble ($s = 1$) can be expressed by

$$\frac{dV_\alpha}{dt} = \frac{4\pi r_\alpha^2}{\left(\frac{dn_\alpha}{dV_\alpha}\right)^2} \frac{D\rho_\beta}{R_g T} \frac{1}{1} \frac{\partial \Delta F}{\partial V_\alpha} \tag{7.15}$$

Substitution of ΔF (eq. (7.4)) into eq. (7.15) yields

$$\frac{dV_\alpha}{dt} = -\frac{4\pi r_\alpha^2}{\left(\frac{dn_\alpha}{dV_\alpha}\right)^2} \frac{D\rho_\beta}{R_g T} \frac{1}{1} \left\{ (p_\beta - p_\alpha + \frac{2\sigma}{r_\alpha}) + (\mu_\alpha - \mu_\beta)\frac{dn_\alpha}{dV_\alpha} \right\} \tag{7.16}$$

and with eq. (7.6) it reads

$$\frac{dr_\alpha}{dt} = -\frac{D\rho_\beta}{\frac{dn_\alpha}{dV_\alpha} R_g T \, 1} (\mu_\beta - \mu_\alpha) \tag{7.17}$$

A truncated Taylor expansion of $\mu_\alpha(p_\alpha)$

$$\mu_\alpha(p_\alpha) = \mu_\alpha(p') + \frac{1}{\rho_\alpha(p')} (p_\beta - p' + \frac{2\sigma}{r_\alpha}) \tag{7.18}$$

is used further to obtain the following expression for the critical radius of the bubble r_c:

$$r_c = \frac{2\sigma}{\rho_\alpha(p')[\mu_\beta(p_\beta) - \mu_\beta(p')] - (p_\beta - p')} \tag{7.19}$$

This is approximately equal to

$$r_c = \frac{2\sigma}{(p'-p_\beta)\left[1 - \frac{\varrho_\alpha(p')}{\varrho_\beta(p')}\right]} \approx \frac{2\sigma}{p'-p_\beta} \tag{7.20}$$

Note, that the necessary condition for the formation of bubbles consists in $p_\beta < p'$.

A substitution of r_c (eq. (7.19)) into the growth equation (7.17) yields:

$$\frac{dr_\alpha}{dt} = \frac{2\sigma D \varrho_\beta}{\varrho(p')\frac{dn_\alpha}{dV_\alpha} R_g T} \frac{1}{l} \left\{\frac{1}{r_c} - \frac{1}{r_\alpha}\right\} \tag{7.21}$$

The necessary condition for eq. (7.21) being in agreement with the results of the thermodynamic investigation consists in $dn_\alpha/dV_\alpha > 0$. Consequently, this condition is fulfilled not only for the equilibrium states (see, eqs. (3.60), (3.61)) but always, if eq. (7.6) holds.

If more than one bubble is present in the system the growth equations for the different bubbles can be obtained in the same way. Instead of eq. (7.15) we get, e.g., for the time evolution of the j-th bubble

$$\frac{dV_\alpha^{(j)}}{dt} = -\frac{4\pi r_\alpha^{(j)2}}{\left(\frac{dn_\alpha^{(j)}}{dV_\alpha^{(j)}}\right)^2} \frac{D\varrho_\beta}{R_g T} \frac{1}{l^{(j)}} \frac{\partial \Delta F}{\partial V_\alpha^{(j)}} \tag{7.22}$$

where ΔF is given by eq. (7.2). This equation can be transformed into

$$\frac{dr_\alpha^{(j)}}{dt} = \frac{2\sigma D \varrho_\beta}{\varrho_\alpha(p')\frac{dn_\alpha^{(j)}}{dV_\alpha^{(j)}} R_g T} \frac{1}{l^{(j)}} \left\{\frac{1}{r_c} - \frac{1}{r_\alpha^{(j)}}\right\} \tag{7.23}$$

In this general case, r_c depends on the number and the size of all bubbles present in the system.

By the same method as discussed in chapter 6.3. the following differential equation for the mean radius of the bubbles is obtained:

$$s\frac{dR}{dt} = -\frac{\omega}{\left(\frac{dn_\alpha}{dV_\alpha}\right)^2} \frac{D\varrho_\beta}{R_g T} \frac{1}{l} \frac{\partial \Delta F}{\partial R} \; ; \quad \frac{\partial \Delta F}{\partial R} = \frac{\partial \Delta F}{\partial s} \frac{ds}{dR} \tag{7.24}$$

ΔF being expressed by eq. (7.8).

Taking into account the condition of mechanical equilibrium (7.6) the relation between the number of bubbles and their mean radius can be determined by

$$\left(\frac{\partial \Delta F}{\partial R}\right)_s = -\, 4\pi R^2 s\left\{ (p_\alpha - p_\beta - \tfrac{2\sigma}{R}) - (\mu_\alpha - \mu_\beta)\,\frac{dn_\alpha}{dV_\alpha}\right\} \tag{7.25}$$

or by the Gibbs-Thomson equation

$$p_\alpha - p_\beta - \frac{2\sigma}{R} - (\mu_\alpha - \mu_\beta)\,\frac{dn_\alpha}{dV_\alpha} = 0 \tag{7.26}$$

The second derivative of ΔF with respect to R reads

$$\left(\frac{\partial^2 \Delta F}{\partial R^2}\right)_s = -\, 8\pi\sigma s\left\{ 1 + Z\left[1 - V_\alpha \frac{d\varrho_\alpha}{dV_\alpha}\frac{1 - s\frac{V_\alpha}{V}}{(\varrho - \varrho_\alpha)}\right] - \frac{3}{2}\frac{V_\alpha^2 R}{\sigma}\frac{\partial \mu_\alpha}{\partial \varrho_\alpha}(\frac{d\varrho_\alpha}{dV_\alpha})^2\right\} \tag{7.27}$$

where Z is obtained as

$$Z = -\, \frac{3R}{2\sigma}\frac{sV_\alpha}{V}\frac{\varrho - \varrho_\alpha}{(1 - s\frac{V_\alpha}{V})^2}\frac{\partial \mu_\beta}{\partial \varrho_\beta}\left(\varrho_\beta - \frac{dn_\alpha}{dV_\alpha}\right) \tag{7.28}$$

In the derivations the relation

$$\frac{dn_\alpha}{dV_\alpha} = \varrho_\alpha + V_\alpha\,\frac{d\varrho_\alpha}{dV_\alpha} \tag{7.29}$$

was used.
A derivation of eq. (7.26) with respect to s results further in

$$\frac{dR}{ds} = -\, \frac{R}{3s}\frac{Z}{\left[1 + Z\left(1 - V_\alpha \frac{d\varrho_\alpha}{dV_\alpha}\frac{1 - sV_\alpha/V}{(\varrho - \varrho_\alpha)}\right) - \frac{3}{2}\frac{V_\alpha^2 R}{\sigma}\frac{\partial \mu_\alpha}{\partial \varrho_\alpha}\left(\frac{\partial \varrho_\alpha}{\partial V_\alpha}\right)^2\right]} \tag{7.30}$$

Together with the eq. (7.24) it yields

$$\frac{dR}{dt} = \frac{\omega\sigma D\varrho_\beta}{\left(\frac{dn_\alpha}{dV_\alpha}\right)^2 R_g T l R}\left\{\frac{1}{Z}\left[1 - \frac{3}{2}\frac{V_\alpha^2 R}{\sigma}\frac{\partial \mu_\alpha}{\partial \varrho_\alpha}\left(\frac{d\varrho_\alpha}{dV_\alpha}\right)^2\right] + 1 - V_\alpha \frac{d\varrho_\alpha}{dV_\alpha}\frac{1 - s\frac{V_\alpha}{V}}{(\varrho - \varrho_\alpha)}\right\} \tag{7.31}$$

with $\omega = 8/27$ (6.46).
The second differential equation for the determination of the number
of bubbles can be obtained from eq. (7.26), again. It reads

$$\frac{d}{dt}\left\{\ln\left(\frac{sV_\alpha}{V}\right)\right\} = -\, \frac{1}{Z}\left\{1 - Z\frac{V_\alpha(1 - s\frac{V_\alpha}{V})}{\varrho - \varrho_\alpha}\frac{d\varrho_\alpha}{dV_\alpha} - \frac{3RV_\alpha^2}{2\sigma}\frac{\partial \mu_\alpha}{\partial \varrho_\alpha}\left(\frac{d\varrho_\alpha}{dV_\alpha}\right)^2\right\}\frac{d\ln R^3}{dt} \tag{7.32}$$

Eqs. (7.31) and (7.32) can be used for the determination of the mean
radius and the number of bubbles in the whole course of the process of
Ostwald ripening along the valley of the thermodynamic potential. Along
this valley the following inequality is fulfilled (see eq. (7.27):

$$1 + Z\left[1 - V_\alpha \frac{1 - s\frac{V_\alpha}{V}}{(\varrho - \varrho_\alpha)} \frac{d\varrho_\alpha}{dV_\alpha}\right] - \frac{3}{2} \frac{V_\alpha^2 R}{\sigma} \frac{\partial \mu_\alpha}{\partial \varrho_\alpha} \left(\frac{d\varrho_\alpha}{dV_\alpha}\right)^2 \leq 0 \qquad (7.33)$$

where the equality sign holds only for the highest point of the valley
of the thermodynamic potential. It follows that also in the case of
bubbles in the first stage of the ripening process R increases relati-
vely slowly with time. Since $Z \sim R^4$, $dn_\alpha/dV_\alpha \sim R^{-4}$ for large mean values
of the size of the bubbles ($dn_\alpha/dV_\alpha \approx \varrho_\alpha$, $\varrho_\beta \approx \varrho'$) we obtain

$$\frac{dR}{dt} = \frac{\omega \sigma D \varrho'}{\varrho_\alpha^2 R_g T} \frac{1}{1R} \quad ; \quad \frac{sV_\alpha}{V} \approx \text{constant} \qquad (7.34)$$

Again, in the asymptotic region the time evolution is described by the
same equations as derived first by Lifshitz and Slyozov.
The outlined results are valid, if the initial supersaturation is not
too high, it means, if the first stage of the transition can be described as a nucleation process. For high initial supersaturations nucleation is replaced by spinodal decompositions which results in a qualitatively different picture of the transition.
It should be underlined, that the results obtained, are widely independent on any specific properties of both phases. Consequently, the
equations can be used also for the investigation, e.g., of the formation and coalescence of vacancies in a solid. As another special case
we get from eqs. (7.31) and (7.32) the equations describing the growth
of drops in a one-component gas under isothermal and isochoric conditions (for $dn_\alpha/dV_\alpha = \varrho_\alpha$).

7.4. Application to Liquid-Gas Solutions and Multicomponent Systems

The results of chapter 6.2. and 6.3. can be generalized to include the
description of the growth and Ostwald ripening of bubbles in liquid-
gas solutions. The crux to such a generalization consists again in the
assumption of an quasi-equilibrium with respect to the pressure. In the
same way, as discussed for bubbles in one-component systems, the growth
equation for one bubble in the liquid-gas solution can be obtained as

$$\frac{dr_\alpha}{dt} = \frac{2 D \varrho_{2\beta} \sigma}{\varrho_\alpha(p)\left(\frac{dn_\alpha}{dV_\alpha}\right) R_g T} \frac{1}{r_\alpha} \left\{\frac{1}{r_c} - \frac{1}{r_\alpha}\right\} \qquad (7.35)$$

r_c being determined by

174

$$r_c = \frac{2\sigma}{\varrho_\alpha(p)\,(\mu_{2\beta} - \mu_\alpha(p))} \qquad (7.36)$$

The time evolution of the mean radius of the bubbles and their number
in the process of Ostwald ripening is governed by the following equations (SCHMELZER, SCHWEITZER, 1987a)

$$\frac{dR}{dt} = \frac{\omega D \varrho_{2\beta} \sigma}{\varrho_\alpha \left(\frac{dn_\alpha}{dV_\alpha}\right) R_g T} \; \frac{1}{R^2} \left\{ \frac{1}{Z} \left(1 - \frac{2\sigma}{3\varrho_\alpha R} \frac{\partial \varrho_\alpha}{\partial p_\alpha} \right) + 1 \right\} \qquad (7.37)$$

$$\frac{d}{dt} \left\{ \ln(sn_\alpha) \right\} = -\frac{1}{Z}\left\{ 1 - \frac{2\sigma}{3\varrho_\alpha R} \frac{\partial \varrho_\alpha}{\partial p_\alpha} \right\} \frac{1}{\varrho_\alpha} \left(\frac{dn_\alpha}{dV_\alpha} \right) \frac{d}{dt}\left\{ \ln(R^3) \right\} \qquad (7.38)$$

$$Z = -\frac{3Rs}{2\sigma \varrho_\alpha} \left(\frac{dn_\alpha}{dV_\alpha} \right)^2 \frac{\partial \mu_{2\beta}}{\partial x_\beta} \frac{n_1 n_\alpha}{(n_1 - n_2 - sn_\alpha)^2} \qquad (7.39)$$

As a special case (for $\partial \varrho_\alpha / \partial p_\alpha = 0$) the eqs. (6.28), (6.36), (6.39)
and (6.40) are obtained again, valid for the description of Ostwald
ripening in solid and liquid solutions.
The approach used here can be applied also for the description of the
formation and growth of clusters of a new phase in multi-component systems. Again, if the steady-state nucleation rate is known, depletion
effects can be taken into account by the quasi-state approximation.

If the system consists of k components the total number of degrees of
freedom for a description of a cluster is equal to $f = k+1$ (for isothermal processes). But if the growth of the clusters is limited mainly
by the diffusion of only one component, denoted by 1, it can be assumed, that at each moment there exists a quasi-equilibrium with respect
to the diffusion of all other components. Thus it yields:

$$p_\alpha - p_\beta = \frac{2\sigma}{r_\alpha} \; ; \quad \mu_{i\alpha} = \mu_{i\beta} \; ; \quad i = 1,2,\ldots,k \qquad i \neq 1 \qquad (7.40)$$

taking into account in addition eq. (7.6).
Therefore the number of degrees of freedom is reduced to one, again,
and eq. (7.14) is applicable, where n_α has to be replaced by $n_{1\alpha}$.
The time evolution of the cluster can be determined then by eqs. (7.14)
and (7.19).
Independent on the specific properties of the system considered the
depletion of the medium surrounding the bubbles or drops of the new
phase leads to the same general scenario of first-order phase transitions.

8. Nucleation and Growth of Clusters in Elastic and Viscoelastic Media

8.1. Derivation of a Growth Equation for Clusters in Elastic Media

Phase transitions in solids, in solid solutions are often accompanied
by the development of elastic strains due to the formation and growth
of the clusters of the new phase. These elastic strains quantitatively
and qualitatively modify the kinetics of growth of single clusters and
of ensembles of clusters, in particular, they modify the kinetics of
Ostwald ripening.

So it was observed in experimental investigations of Ostwald ripening
of AgCl clusters in a sodium borate melt, that elastic strains may lead
to a qualitative different compared with the results of Lifshitz and
Slyozov behaviour for the evolution of the mean cluster radius and the
number of clusters (PASCOVA, GUTZOW, 1983; GUTZOW, SCHMELZER, PASCOVA,
POPOW, 1985). Instead of a linear growth of the mean volume and a decrease of the number of clusters an asymptotic behaviour as shown in
Fig. 8.1 was found.

For a special type of growth equation of the clusters such a behaviour
was described based on a modification of the theory of Lifshitz and
Slyozov (GUTZOW et al., 1985). But because of the great variety of
practical applications of such processes and the variety of different
types of strains developing a general method of the kinetic description
of the growth of single clusters and ensembles of clusters in elastic
and viscoelastic media is of importance. This general theoretical
approach is based on the growth equation (5.21) and on the outlined
in chapter 6 new theory of the process of Ostwald ripening.

Using the diffusion equation approach (Chapter 5.3.) the influence of
elastic strains on Ostwald ripening can be expressed by the introduction of an additional term $\mu_i^{(\varepsilon)}$ to the chemical potential, which describes the influence of elastic strains on the diffusion process. So,
instead of eq. (5.5) we can write

$$\vec{j}_i = - \frac{D_i c_i}{kT} \, \text{grad}(\mu_i + \mu_i^{(\varepsilon)}) \tag{8.1}$$

To find from eq. (8.1) the growth equation for a single cluster the
quantity grad $\mu(r)$ has to be known for $r = r_\alpha$, the radius of the
cluster.

The general structure of $\mu(r)$ can be suggested without serious difficulties. Since in the spherical symmetric case all quantities describing the elastic deformation are functions of the form $f(r_\alpha/r)$, it is
reasonable to assume, that $\mu^{(\varepsilon)}(r)$ can be expressed in the same way as

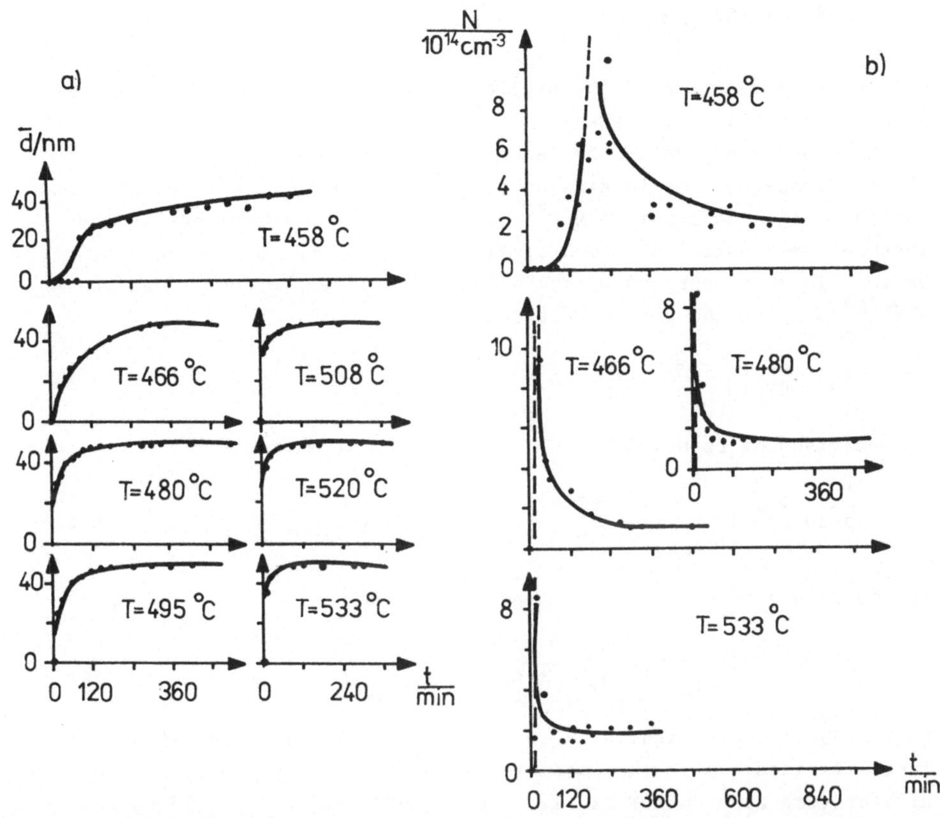

Fig. 8.1:
a) Mean diameter $\bar{d}$ (in nm) of AgCl-clusters as a function of time as
   observed experimentally for the process of Ostwald ripening of
   AgCl-clusters in a sodium borate melt at different temperatures
   indicated at each curve (PASCOVA, GUTZOW, 1983; GUTZOW et al.,
   1985). The authors found a practically stationary asymptotic behaviour corresponding to a mean cluster radius of about 20 nm with
   a relatively monodisperse distribution with respect to cluster sizes.
b) Number of clusters N (in 10^{14} cm^{-3}) as a function of time for formation (increasing initial part of each curve) and ripening process of AgCl phase in sodium borate melt. There is found a practically stationary value of about 10^{14} cm^{-3} for the number of
   clusters.

(From: PASCOVA, GUTZOW, 1983)

$$\mu^{(\varepsilon)} = \mu^{(\varepsilon)} \left(\frac{r_\alpha}{r} \right) \tag{8.2}$$

But the determination of the explicite form of this dependence is, in general, a complicated task.

On the other hand, using the general growth equation (5.20) or (5.21) the influence of elastic strains can be taken into account by an calculation of the contribution $\Phi^{(\varepsilon)}$ of the elastic deformations to the thermodynamic potential. Such a calculation can be carried out much easier. In this approach we get an additional term proportional to $\partial \Phi^{(\varepsilon)} / \partial n_\alpha$ in the growth equation:

$$j_R = \frac{Dc}{kT} \frac{1}{l} \left\{ \frac{\partial \Phi}{\partial n} + \frac{\partial \Phi^{(\varepsilon)}}{\partial n} \right\} \tag{8.3}$$

A comparison of both methods shows, that the term grad $\mu^{(\varepsilon)}$ is replaced by

$$\left. \mathrm{grad}\; \mu^{(\varepsilon)} \right|_{r=r_\alpha} = \frac{1}{l} \frac{\partial \Phi^{(\varepsilon)}}{\partial n} \tag{8.4}$$

Taking into account eq. (8.2) grad $\mu^{(\varepsilon)}$ reads

$$\left. \mathrm{grad}\; \mu^{(\varepsilon)} \right|_{r=r_\alpha} = - \frac{\bar{r}_\alpha}{r_\alpha^2} \left. \frac{\partial \mu^{(\varepsilon)}}{\partial (r_\alpha / r)} \right|_{r=r_\alpha} \tag{8.5}$$

Eq. (8.5) indicates that also in the case of elastic strains for diffusion limited growth the parameter l is equal to the radius of the cluster (see also GUTZOW et al., 1985; SCHMELZER, GUTZOW, 1986a,b).

Again, a generalization of this approach to non-spherical interfaces is straighforward.

8.2. Models for the Calculation of Elastic Strains

For an application of eq. (8.2) the contribution of the elastic strains $\Phi^{(\varepsilon)}$ to the thermodynamic potential Φ (the total energy of elastic deformations due to the formation of a cluster) has to be known. This elastic energy will be calculated here for three different models (see also: LANDAU, LIFSHITZ, 1977; TIMOSCHENKO, GOODIER, 1975).

8.2.1. Elastic Strains of Nabarro Type

Let us assume first, that in the course of the phase transition (allotropic transition or crystallization process) inside an approximately
spherical body with the radius R_2 and the elastic constants E_2 (Young's
modulus) and γ_2 (Poisson's number) a spherical nucleus of a new phase
with the elastic constants E_1 and γ_1 is formed. The volume per particle in the absence of elastic strains is equal to v_2 and v_1, respectively. In the absence of elastic strains the radius of the cluster
would be equal to R instead of R' in the deformed state. So the difference between the volume per particle in both phases results in the
development of elastic strains (see Fig. 8.2).
The model developed here gives the possibility to analyze the influence of finite size effects on the development of elastic strains.
In the limit $R_2 \rightarrow \infty$ as a special case we obtain the energy of deformation for a cluster in an infinite matrix.
Since the total number of particles in the cluster can be expressed by
$z = 4\pi R^3/3v_1 = V/v_1$ the difference in the total volume of the z particle in both phases is equal to

$$V_t = z(v_2-v_1) = V(v_2-v_1)/v_1 \tag{8.6}$$

Defining a quantity $\delta = (v_2-v_1)/v_1$ we can also write $\quad V_t = \delta V.$

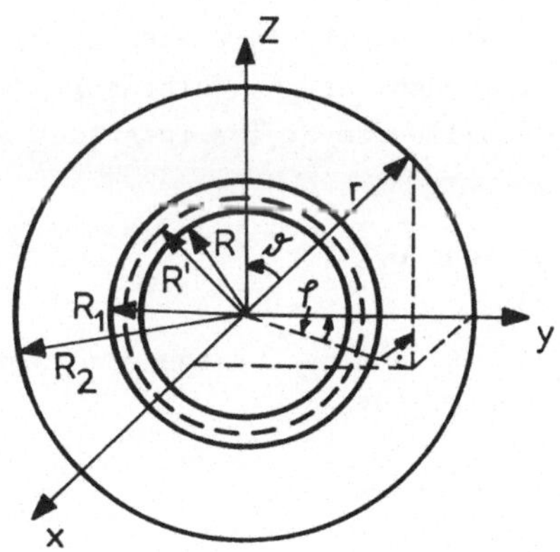

Fig. 8.2:
General model of a nucleus in a matrix used for the calculation of
elastic strains in allotropic transitions or crystallization processes
R_1,R_2 - inner and outer radius of the hollow sphere (matrix) in the
            undeformed state
R   - radius of the cluster in the undeformed state
R' - inner radius of the matrix and radius of the cluster in the de-
            formed state (From: HOELL, SCHMELZER, MAHNKE, 1986)

If the cluster and the surrounding medium remain in close contact
(as it is assumed here) the components of the stress tensor and the
tensor of deformations for the cluster are given by (SCHMELZER,
1985a; HOELL, 1986; see also HEINRICH, ULBRICHT, 1981)

$$u_r = C_1 r \qquad \text{with} \qquad C_1 = \frac{1}{3}\frac{\Delta V}{V}$$

$$u_{rr} = u_{\theta\theta} = u_{\varphi\varphi} = C_1 \; ; \quad \sigma_{rr} = \sigma_{\theta\theta} = \sigma_{\varphi\varphi} = C_1 \frac{E_1}{1-2\gamma_1}$$

(8.7)

and for the matrix by

$$u_r = C_1 r + C_2/r^2 \qquad \text{with}$$

$$C_1 = \frac{2(1-2\gamma_2)R_1^2\,\Delta R_1}{(1+\gamma_2)R_2^3 + 2(1-2\gamma_2)R_1^3} \quad ; \quad C_2 = \frac{R_2^2(1+\gamma_2)R_1^2\,\Delta R_1}{(1+\gamma_2)R_2^2 + 2(1-2\gamma_2)R_1^3}$$

(8.8)

$$u_{rr} = C_1 - 2C_2/r^3 \; ; \; u_{\theta\theta} = u_{\varphi\varphi} = C_1 + C_2/r^3$$

$$\sigma_{rr} = C_1 \frac{E_2}{1-2\gamma_2} - C_2 \frac{2E_2}{1+\gamma_2}\frac{1}{r^3}$$

$$\sigma_{\theta\theta} = \sigma_{\varphi\varphi} = C_1 \frac{E_2}{1-2\gamma_2} + \frac{E_2}{1+\gamma_2}\frac{1}{r^3}$$

In eqs. (8.8) the boundary conditions $\sigma_{rr}(R_2) = 0$, $u_r(R_1) = \Delta R_1$
were used. u_r is the radial component of the deformation vector.
The condition of mechanical equilibrium at the interface between cluster and matrix reads

$$\sigma_{rr}^{\text{cluster}}(R') = \sigma_{rr}^{\text{hollow sphere}}(R')$$

(8.9)

Denoting further by $\Delta R_1 = R' - R_1$ and by ΔV the change of the volume of the cluster ΔV_t can be expressed as

$$\Delta V - 4\pi R_1^2\,\Delta R_1 = \delta V$$

(8.10)

Eqs. (8.9) and (8.10) result in

$$\frac{\Delta V}{V} = \frac{6\delta}{B}(1-2\gamma_1)E_2(R_2^3 - R_1^3)$$

$$\Delta R_1 = -\frac{\delta}{B}E_1 R_1\left[(1+\gamma_2)R_2^3 + 2(1-2\gamma_2)R_1^3\right]$$

(8.11)

$$B = 3\left\{(1-2\gamma_1)\left[2E_2(R_2^3-R_1^3)\right] + E_1\left[(1+\gamma_2)R_2^3 + 2R_1^3(1-2\gamma_2)\right]\right\}$$

The density of energy of elastic deformations in both phases is given then by

$$f^{cluster} = 6\; E_1 E_2^2 (1-2\,\gamma_1)(R_2^3 - R_1^3)^2 \frac{\delta^2}{B^2} \qquad (8.12)$$

$$f^{medium} = 6(1-2\,\gamma_2) E_1^2 E_2 R_1^6 \frac{\delta^2}{B^2}\left\{ 1 + \frac{R_2^6}{r^6}\frac{(1+\gamma_2)}{2(1-2\,\gamma_2)}\right\} \qquad (8.13)$$

Integration over the volume yields

$$F^{cluster} = 6\; E_1 E_2^2 (1-2\,\gamma_1)(R_2^3 - R_1^3)^2 \frac{\delta^2}{B^2}\,V \qquad (8.14)$$

$$F^{matrix} = 6(1-2\,\gamma_2) E_1^2 E_2 R_1^3 \left\{ R_2^3 - R_1^3 \; - \; \frac{1+\gamma_2}{2(1-2\,\gamma_2)} \; *\right.$$

$$\left. *\; R_2^6 \left[\frac{1}{R_2^3} - \frac{1}{R_1^3} \right] \right\}\frac{\delta^2}{B^2}\,V \qquad (8.15)$$

In the limit $R_2 \longrightarrow \infty$ this expressions are reduced to

$$F^{cluster} = \frac{2}{3}\;\frac{E_1 E_2^2 (1-2\,\gamma_1)}{\left[2E_2(1-2\,\gamma_1)+E_1(1+\gamma_2)\right]^2}\;\delta^2 V \qquad (8.16)$$

$$F^{matrix} = \frac{E_2 E_1^2 (1+\gamma_2)}{3\left[2E_2(1-2\,\gamma_1)+E_1(1+\gamma_2)\right]^2}\;\delta^2 V \qquad (8.17)$$

and tho total energy of elastic deformations can be expressed by

$$F = F^{cluster} + F^{matrix} = \frac{E_1 E_2}{3\left[2E_2(1-2\,\gamma_1)+E_1(1+\gamma_2)\right]}\;\delta^2 V \qquad (8.18)$$

This limiting case is reached in a good approximation for $R_2 \gtrsim 5R_1$.
If matrix and cluster have the same elastic constants eq. (8.18) gets the simple form

$$F = \frac{E}{9(1-\gamma)}\;\delta^2 V \qquad (8.19)$$

an expression obtained already by NABARRO (1940). Elastic strains of the type (8.19) we denote therefore as strains of Nabarro-type (Fig. 8.3).

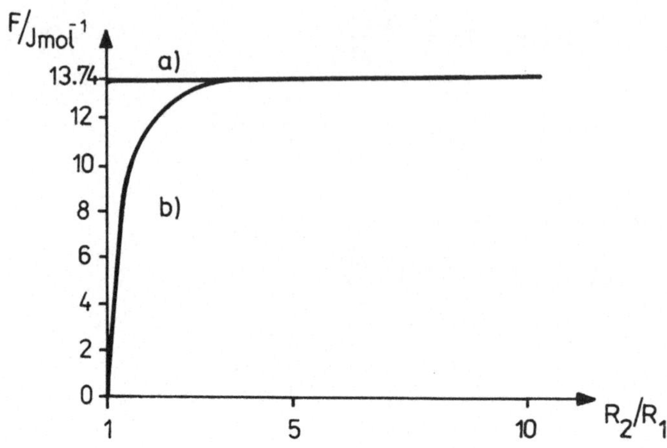

Fig. 8.3:
a) Elastic energy of Nabarro-type (8.19) in the limit $R_2 \to \infty$
b) Elastic energy following eqs. (8.14; 8.15) with $E = E_1 = E_2$; $\gamma = \gamma_1 = \gamma_2$ for finite R_2
Calculations were done for the system β-cristobalite-silica-glass with $E = 7 \cdot 10^{10}$ Pa, $\gamma = 0.17$; $\delta = 7.3 \cdot 10^{-3}$; $V = 27.519 \cdot 10^{-6}$ m³.
(From: HOELL, 1986)

If the volume of the cluster is expressed, again, by V_α , we come to
the conclusion, that elastic strains due to differences of the volume
per particle in the different phases lead to contributions of the
form

$$\Phi^{(\varepsilon)} = \varepsilon \, V_\alpha \qquad \varepsilon = \text{const.} \tag{8.20}$$

to the characteristic thermodynamic potential, if the density of
distribution of the clusters in the system is not too high (the distance between the clusters must be greater than ten times the radius
of the cluster, in this case the eqs. (8.18), (8.19) are valid.) Note
further, that since $F \sim \delta^2$, the results of the calculations do not
depend on the sign of δ .

8.2.2. Elastic Strains in Segregation Processes in Elastic Media

In a solid solution under certain conditions the formation of clusters
of a new phase due to segregation processes of, e.g., one of the components is possible. In the first stage of growth of the clusters this
process is due to the direct exchange of places of the particles of
the different components. Elastic strains evolving in this stage can
be described by eq. (8.20), again (see, e.g., CHRISTIAN, 1975).

After a certain volume V_o is reached the further growth of the
cluster is due to a transport by diffusion of the segregating particles over finite distances. As a result in the surrounding the cluster
matrix the elastic strains grow more rapidly.

If one assumes, that the matrix can be considered as an elastic
(Hooke's) body the evolving strains can be calculated in the following way. The growth of the cluster leads to an increase of the inner
radius of the matrix expressed by ΔR_1. The coefficients C_1 and C_2
characterizing the state of elastic deformation of the matrix can be
expressed then by $C_1 = 0$, $C_2 = R_o^2 \Delta R_1$. For simplicity of the derivations the case of an infinite matrix is considered, R_o is the radius
corresponding to the volume V_o, where the new mechanism of development
of the elastic strains becomes determining.

From the equilibrium conditions at the interface between the cluster
and the matrix we obtain for the resulting pressure on the cluster

$$p = \frac{2}{3} \frac{E_2}{1+\gamma_2} \frac{\Delta V}{V_o} \tag{8.21}$$

The densities of the energy of deformations in both phases are given
then by

$$f^{cluster} = \frac{2}{3} \frac{E_2^2(1-2\gamma_1)}{E_1(1+\gamma_2)^2} \left(\frac{\Delta V}{V_o}\right)^2 \tag{8.22}$$

$$f^{matrix} = \frac{3E_2}{1+\gamma_2} \frac{(\Delta V)^2}{4\pi r^3}$$

resulting in the following expression for the total energy of elastic
deformations

$$F = \frac{1}{3} \frac{E_2}{1+\gamma_2} \left\{ \frac{2E_2(1-2\gamma_1)+E_1(1+\gamma_2)}{E_1(1+\gamma_2)} \right\} \frac{(\Delta V)^2}{V_o} \tag{8.23}$$

If, in particular, the elastic constants of matrix and cluster are
the same, eq. (8.23) is reduced to

$$F = \frac{E(1-\gamma)}{(1+\gamma)^2} \frac{(\Delta V)^2}{V_0} \tag{8.24}$$

Denoting now the volume of the cluster by V_α, again, we can write

$$\Phi^{(\varepsilon)} = \mathcal{H} \frac{(V_\alpha - V_0)^2}{V_0} \theta \qquad \theta = \begin{cases} 1 & V_\alpha \geqslant V_0 \\ 0 & V_\alpha < V_0 \end{cases} \tag{8.25}$$

8.2.3. The Influence of Viscous Properties of the Matrix on the Development of Elastic Strains

Though equations of the type (8.20) have been derived for the description of strains evolving in allotropic transitions or crystallization processes they are also applied widely to the calculation of
elastic strains in solid solutions developing as a result of an exchange of the places of the particles of different components
(CHRISTIAN, 1975). In the last chapter it has been shown, however,
that such an application is not always correct. If the particles of
the second component (matrix) move very slowly (their diffusion coefficient D_M tends to zero) then we have to deal in this limit with a
segregation process in a purely elastic medium leading instead of eq.
(8.20) to expressions of the typ (8.25).
Further, if D_M is much greater than the diffusion coefficient D of
the segregating particles no elastic strains will evolve at all. Consequently, eq. (8.20) in application to solid solutions is obviously
valid only for $D_M \simeq D$, eq. (8.25) in the limit $D_M \rightarrow 0$
If the diffusion coefficient D_M of the matrix particles has a finite
value in the intervall $0 < D_M < D$, the matrix has both elastic and
viscous properties and elastic strains will relax with time. Since
such relaxation effects are of importance for different applications,
e.g., crystallization processes in glassforming melts, we will develop here a special model taking into account such phenomena. According to the discussion given above, this model is applicable in the
interval $0 < D_M/D < 1$.
If the volume of the cluster is less than V_0 the elastic strains can
be expressed by $\Phi = \varepsilon V_\alpha$. Since it is assumed, that for $V < V_0$ elastic strains evolve due to an exchange of the particles of the diffe-
rent components, the minimum value of the width of the inhomogeneous
region with an increased number of matrix particles and a decreased

number of segregating particles is given by

$$\frac{4\pi}{3} R_0^3 c_\alpha = \frac{4\pi}{3} (R_0 + d_0)^3 c \qquad (8.26)$$

c is the concentration of the segregating particles in the homogeneous
initial state, c_α is the concentration in the cluster.
Starting with the volume $V = V_0$ the transport of the segregating particles over finite distances to the growing cluster results in an
additional increase of the density of matrix particles in its neigh-
bourhood. This additional contribution to the density will depend on
the distance to the cluster as shown in Fig. 8.4.
The real distribution of additional matrix particles in the model is
replaced by a distribution of the form $\Delta\varrho$ = constant in the interval
$R < r < R+d$, $\Delta\varrho = 0$ otherwise. The constant is determined in such a
way, that the total number of additional matrix particles is the same
in the real and in the idealized distribution.
If diffusion processes of the matrix particles can be neglected the
parameter d is equal, approximately, to $d = d_0-(R-R_0)$. Diffusion processes lead to an increase of d which according to the diffusion laws
can be expressed by a term $\sqrt{D_M t}$. So we arrive, finally, at

$$d = d_0 - (R - R_0) + \sqrt{D_M t} \qquad (8.27)$$

The formation of a region with an increased density of matrix particles
results in an increase of the work, necessary for an introduction of
additional segregating particles to the cluster. Mathematically we can
express this fact by the ansatz (8.28) for $\phi^{(\varepsilon)}$

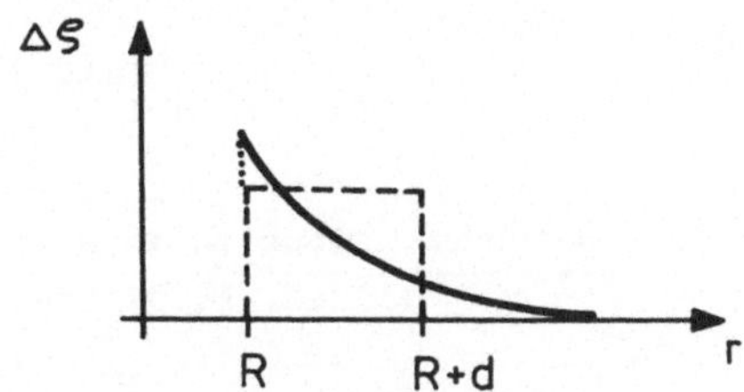

Fig. 8.4:
Real (———) and idealized (-----) distribution of additional matrix
particles around the growing cluster. The width of the inhomogeneous
region as assumed in the model is determined by eq. (8.27).

(From: SCHMELZER, 1985a)

$$\phi(\varepsilon) = \begin{cases} \varepsilon V & V \leq V_0 \\ \varepsilon V_0 + (V - V_0)\, \pi(\Delta\varrho) & V > V_0 \end{cases} \tag{8.28}$$

$\pi(\Delta\varrho)$ is a (unknown) function of the excess density $\Delta\varrho$ of matrix particles in the environment of the cluster. Since the volume of the inhomogeneous region around the cluster is given by

$$V_{in} = \frac{4\pi}{3}\left[(R + d)^3 - R^3\right] \tag{8.29}$$

the number of additional matrix particles by

$$z = \frac{4\pi}{3}(R^3 - R_0^3)/v \tag{8.30}$$

where v is the volume of one matrix particle in the initial state, the excess density $\Delta\varrho$ can be expressed by

$$\Delta\varrho = \frac{R^3 - R_0^3}{v((R + d)^3 - R^3)} \tag{8.31}$$

A truncated Taylor expansion of the function $\pi(\Delta\varrho)$ leads to $\pi(\Delta\varrho) = \pi_0 + \pi_1 \Delta\varrho + \ldots$ π_0 and π_1 are constants, which are determined from the two limiting conditions.

For $V \leq V_0$ or $D_M \simeq D$ $\Delta\varrho = 0$ and $\phi^{(\varepsilon)}$ is equal to $\phi^{(\varepsilon)} = \varepsilon V$. Consequently, π_0 is equal to ε . In the second limiting case ($D_M = 0$) the second term is much greater than the first and moreover, $R \simeq R_0$, $d \simeq d_0$. So we get in this limiting case from eq. (8.28)

$$\phi^{(\varepsilon)} \simeq \frac{\pi_1 (V - V_0)^2}{v\, V_0\left[(1 + \frac{d_0}{R_0})^3 - 1\right]} \tag{8.32}$$

which has to be equivalent to eq. (8.25). Moreover, since in this limiting case d_0 can be expressed by $d_0 = R_0(\sqrt[3]{c_\alpha/c} - 1)$ eq. (8.32) reads

$$\phi^{(\varepsilon)} = \frac{\pi_1 (V - V_0)^2}{v\, V_0\left[c_\alpha/c - 1\right]} \tag{8.33}$$

A comparison with eq. (8.25) shows, that π_1 is equal to

$$\pi_1 = \varkappa\, v\left[\frac{c_\alpha}{c} - 1\right] \tag{8.34}$$

Consequently, we obtain finally

$$\phi(\varepsilon) = \varepsilon V + \varkappa' \frac{(V - V_0)^2}{V} \tag{8.35}$$

$$\varkappa' = \frac{\varkappa(c_\alpha/c - 1)}{\left[1 + \dfrac{R_0(\sqrt[3]{\frac{c}{c}} - 1) + \sqrt{D_M t}}{R} \right]^3 - 1} \tag{8.36}$$

If $D_M \simeq 0$ then $R \simeq R_0$ and $\varkappa'$ equals $\varkappa$. In contrast, if the first term in the denominator of eq. (8.36) is much greater compared with one, $\varkappa'$ is nearly equal to

$$\varkappa' = \frac{\varkappa(c_\alpha/c - 1)}{\left[\dfrac{R_0(\sqrt[3]{c_\alpha/c} - 1) + \sqrt{D_M t}}{R} \right]^3} \tag{8.37}$$

A substitution of this value into eq. (8.35) leads to the same result as if we replace the eqs. (8.35) and (8.36) by

$$\phi(\varepsilon) = \varepsilon V + \varkappa'' \frac{(V - V_0)^2}{V_0} \theta \ , \tag{8.38}$$

$$\varkappa'' = \frac{\varkappa(c_\alpha/c - 1)}{\left[1 + \dfrac{R_0(\sqrt[3]{c_\alpha/c} - 1) + \sqrt{D_M t}}{R_0} \right]^3 - 1}$$

Since in both limiting cases the eqs. (8.35), (8.36) and (8.38) are
equivalent, they lead to nearly equivalent results also in the inter
mediate region. Since eqs. (8.38) are considerably easier to use in
applications we will start with them in the investigation of growth of
clusters in viscoelastic media.
Such a simplification is in addition motivated by the argument, that
the outlined model contains a number of assumptions, approximations, so
that only a qualitative exact description can be awaited. Qualitatively, the kinetics of growth will be the same for both expressions
(8.35), (8.36) and (8.38), respectively.
Despite the mentioned limitations of the model outlined, it seems to
us, that the model has advantages compared with a single statement of
a special type of rheological behaviour, since the mechanism underlying
the viscous behaviour is reflected clearly and only three constants
describing the properties of the matrix are used (E, γ, D_M) which can

be determined, in principle, by independent of the model measurements.
A fourth parameter V_0 can be calculated based on a more detailed microscopic model of the formation and growth of the cluster. At the present
stage of development of the theory it is the only adjustable parameter.

Denoting the volume of the cluster by V_α , again, we obtain

$$\phi^{(\varepsilon)} = \varepsilon V_\alpha + \varkappa'' \frac{(V_\alpha - V_0)^2}{V_0} \theta \tag{8.39}$$

where $\varkappa'$ is determined by eq. (8.38).

8.3. Formation and Growth of Single Clusters in Elastic Media

In the discussion of the influence of elastic strains on the process
of formation and growth of the clusters of a new phase first a possible
depletion is neglected. Starting with eq . (5.21) and adding to eq.
(5.22) the term $\phi^{(\varepsilon)}$ describing the elastic effects, we obtain

$$\frac{dr_\alpha}{dt} = - \frac{Dc}{c_\alpha kT} \frac{1}{l} \frac{\partial \Delta G}{\partial n_\alpha} \quad , \quad \Delta G = n_\alpha(\mu_\alpha - \mu_\beta) + \sigma A + \phi^{(\varepsilon)} \tag{8.40}$$

or

$$\frac{dr_\alpha}{dt} = \frac{Dc}{c_\alpha kT} \frac{1}{l} \left\{ \mu_\beta(p,x) - \mu_\alpha(p) - \frac{2\sigma}{r_\alpha} - \frac{1}{c_\alpha} \frac{\partial \phi^{(\varepsilon)}}{\partial V_\alpha} \right\} \tag{8.41}$$

The critical cluster radius is determined by

$$r_c = \frac{2\sigma}{\mu_\beta(p,x) - (\mu_\alpha(p) + \frac{1}{c_\alpha} \frac{\partial \phi^{(\varepsilon)}}{\partial V_\alpha})} \tag{8.42}$$

Since in all considered cases $\partial \phi / \partial V_\alpha > 0$, the influence of elastic
strains leads to an increase of the critical cluster size compared with
the case $\phi^{(\varepsilon)} = 0$. In the special, and in the stage of nucleation,
presumably, most important, case $\phi^{(\varepsilon)} = \varepsilon V_\alpha$ eq. (8.42) yields

$$r_c = \frac{2\sigma}{\mu_\beta - (\mu_\alpha + \frac{\varepsilon}{c_\alpha})} \tag{8.43}$$

Since relatively small variations of the critical radius of the cluster
lead to significant variations in the nucleation rate, the existence
of elastic strains strongly influences the process of nucleation, it
may prevent nucleation at all (Fig. 8.5).

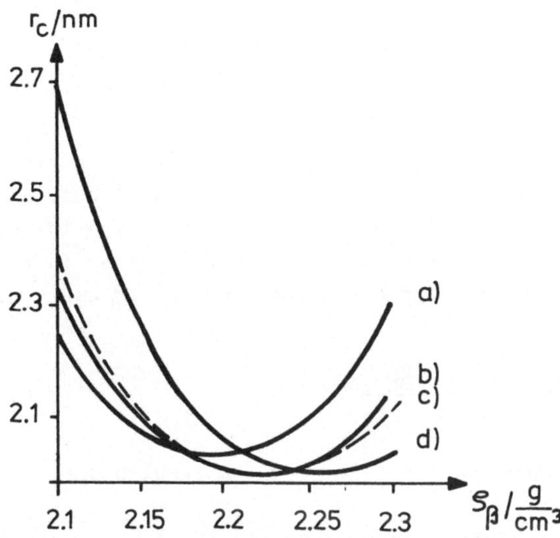

Fig. 8.5:
Critical cluster radius as a function of the matrix density corresponding to different additional hydrostatical pressures p.
a) p = 5 · 10^8 Pa   b) p = 1 · 10^8 Pa
c) p = 0   d) p = - 5·10^8 Pa

(From: HOELL, 1986)

Considering the growth of already formed supercritical clusters two
different cases can be distinguished. If $\phi^{(\varepsilon)} = \dot{\varepsilon}\, V_\alpha$ the strains decrease the growth rate, but once the process of growth has started,
strains of such a type cannot lead to a stop of the growth. On the
other hand, if the elastic strains evolving due to the formation and
growth of the clusters grow more rapidly than linear with the volume
of the cluster, then elastic strains always stop the growth process
for some value of the volume of the cluster. For example, if the strains
are expressed by eq. (8.25) the growth equation reads

$$\frac{dr_\alpha}{dt} = \frac{Dc}{c_\alpha kT}\,\frac{1}{l}\left\{\mu_\beta(p,x) - \left[\mu_\alpha(p) + \frac{2\varkappa}{c_\alpha}\,\frac{V_\alpha - V_0}{V_0}\,\theta\right]\right\} \qquad (8.44)$$

(surface effects are neglected here). The asymptotic value of the volume of the cluster is given then by

$$\mu_\beta(p,x) - \mu_\alpha(p) = \frac{2\varkappa}{c_\alpha}\,\frac{V_\alpha - V_0}{V_0} \qquad (8.45)$$

If in addition to elastic strains viscous properties are taken into
account, then applying, e.g., eq. (8.39), the growth equation is given
by

$$\frac{dr_\alpha}{dt} = \frac{Dc}{c_\alpha kT} \frac{1}{l} \left\{ \mu_\beta - \left[\mu_\alpha + \frac{\varepsilon}{c_\alpha} + \frac{2\varkappa''}{c_\alpha} \frac{V_\alpha - V_0}{V_0} \theta \right] \right\} \tag{8.46}$$

Two stages of growth can be distinguished here, a first stage of rapid
growth, where the difference in the chemical potentials of both phases
is the dominating factor of growth and a second stage, where the growth
rate is determined by the relaxation of the elastic strains. In this
stage $dr_\alpha/dt \simeq 0$ and the time evolution of the cluster radius is given,
approximately, by

$$\mu_\beta(p,x) - (\mu_\alpha(p) + \frac{\varepsilon}{c_\alpha}) = \frac{2\varkappa''}{c_\alpha} \frac{V_\alpha - V_0}{V_0} \tag{8.47}$$

Such a behaviour is analogous to the corresponding results obtained for
a Maxwell body.

Results of numerical calculations based on eq. (8.46) for different
values of the diffusion coefficient D_M of the matrix particles are presented in Fig. 8.6. For high values of D_M the second term $\varkappa''(V_\alpha - V_0)/V_0$
is negligible, for small values of D_M it determines the time evolution
of the cluster size.

8.4. Ostwald Ripening in Elastic and Viscoelastic Media

By the method outlined in chapters 6.2. and 6.3. the following differential equations for the mean radius R and the number of clusters s
in the process of Ostwald ripening can be obtained (see also:
SCHMELZER, GUTZOW, 1986a,b):

$$\frac{dR}{dt} = \frac{c_\beta D \omega}{c_\alpha^2 kT1R} \left\{ \sigma + \frac{3}{4\pi R^2} \left[\phi^{(\varepsilon)} - V_\alpha \frac{\partial \phi^{(\varepsilon)}}{\partial V_\alpha} \right] \right\} \frac{1}{Z} \left\{ 1+Z - \frac{R^2}{2\sigma} \frac{\partial^2 \phi^{(\varepsilon)}}{\partial R \partial V_\alpha} \right\} \tag{8.48}$$

$$\frac{d}{dt} \left\{ \ln(\frac{sV_\alpha}{V}) \right\} = - \frac{1}{Z} \left\{ 1 - \frac{R^2}{2\sigma} \frac{\partial^2 \phi^{(\varepsilon)}}{\partial R \partial V_\alpha} \right\} \frac{d}{dt} \left\{ \ln (R^3) \right\}$$

where Z is determined by eq. (6.28). These equations are analyzed now
for different special cases given by eqs. (8.20), (8.25) and (8.39),
respectively.

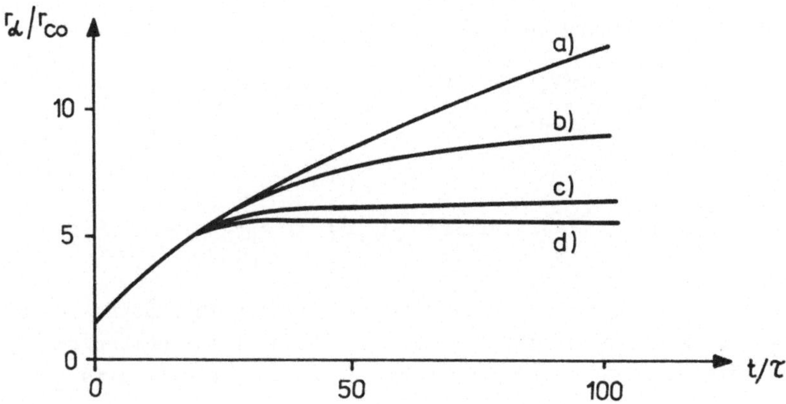

Fig. 8.6:
Growth of a cluster in a visco-elastic medium. For different values
a) - d) of the diffusion coefficient of the matrix particles different
curves are obtained, these curves are bounded by two limiting curves
resulting from

$$\phi^{(\varepsilon)} = \varepsilon\, V_\alpha \quad \text{and} \quad \phi^{(\varepsilon)} = \varepsilon\, V_\alpha + \mathcal{x}\, \frac{(V_\alpha - V_0)^2}{V_0}\, \Theta$$

respectively. The following values of the parameters are used:
a) $D_M = 4.3 \cdot 10^{-15}$ m^2/s; b) $D_M = 4.3 \cdot 10^{-16}$ m^2/s;
c) $D_M = 4.3 \cdot 10^{-17}$ m^2/s; d) $D_M = 4.3 \cdot 10^{-22}$ m^2/s;
$\varepsilon = 2.37 \cdot 10^{-7}$ N/m^2; $\mathcal{x} = 4.8 \cdot 10^{8}$ N/m^2
$\tau - c^2 kT\, r_{co}^2 \mathcal{x} ? \sigma cD) = 0.016$ s
For the values of the other parameters see the caption to Fig.6.1
(From: SCHMELZER, 1985a)

If the elastic strains are expressed by eq. (8.20) (elastic strains
of Nabarro-type) then the eqs.(8.48) are reduced to eqs. (6.39) and
(6.40), again. Therefore, elastic strains of Nabarro-type do not
change qualitatively the kinetics of Ostwald ripening and modify only
quantitatively the relation between the number of clusters s and their
mean radius R.
Since the generalized Gibbs-Thomson equation in the considered case
is given by

$$\ln\left\{\frac{c}{c'}\,\frac{(1-s\,n_\alpha/n)}{(1-s\,V_\alpha/V)}\right\} + \frac{s\,V_\alpha/V}{1-s\,V_\alpha/V} - \frac{2\sigma}{c_\alpha kTR} - \frac{\varepsilon}{c_\alpha kT} = 0 \qquad (8.49)$$

for a given number of clusters the critical radius R_c is greater and
the radius R_{st} less than for the case $\varepsilon = 0$.
If segregation in purely elastic media is considered $\phi^{(\varepsilon)}$ can be expressed by eq. (8.25) and eqs. (8.48) yield

$$\frac{dR}{dt} = \frac{\omega c_\beta D}{c_\alpha^2 kT1R} \left\{ \sigma - \frac{3\varkappa(V_\alpha^2 - V_0^2)}{4\pi R^2 V_0}\theta \right\} \frac{1}{Z} \left\{ 1+Z-3 \frac{\varkappa V_\alpha R\theta}{\sigma V_0} \right\}$$

$$(8.50)$$

$$\frac{d}{dt}\left\{ \ln\left(\frac{sV_\alpha}{V}\right) \right\} = -\frac{1}{Z}\left\{ 1 - \frac{4\pi R^2 \varkappa}{\sigma V_0}\theta \right\}\frac{d}{dt}\left\{ \ln(R^3) \right\}$$

In this case along the valley of the characteristic thermodynamic potential there exists a minimum for finite values of the number of
clusters. This minimum is determined by eq. (6.31) resulting in

$$\frac{d\Delta G}{ds} = \frac{1}{3}\sigma A - \varkappa \frac{V_\alpha^2 - V_0^2}{V_0}\theta = 0 \qquad (8.51)$$

It corresponds to a stable thermodynamic equilibrium state the system
of clusters tends to widely independent of the initial conditions.
In this asymptotic state the distribution of clusters is nearly monodisperse, since the limiting the growth of the clusters factor - the
elastic strains - are of the same type for all of the clusters.

In Fig. 8.7 results of numerical integrations of the eqs. (8.50)
are presented. In the first stage ($V_\alpha \lesssim V_0$) elastic strains modify the
behaviour only quantitatively. This is expressed, e.g., in the some-
what modified initial state of the ripening process, which corresponds,
now, to a mean radius $r_{\alpha c}$ = 9.05 10^{-10} m and a number of clusters s_c
= 1.26 $\cdot 10^{24}$ m^{-3} compared with the values $r_{\alpha c}$ = 8.25 $\cdot 10^{-10}$ m and
s_c = 1.84 $\cdot 10^{24}$ m^{-3} obtained for $\varepsilon = 0$ (see Fig. 6.1). In the second
stage ($V_\alpha > V_0$) strains of the type (8.25) become more and more important and lead to a rapid decrease of the growth rate and a stationary
asymptotic behaviour.
Such a type of evolution of the ensemble of clusters is always to be
expected, if the elastic strains grow more rapidly as linear with the
volume of the cluster, since according to eq. (6.31) only in these
cases there exists a minimum of the thermodynamic potential for finite
values of the number of clusters.
This conclusion is in contrast to the opinion expressed by LIFSHITZ
and SLYOZOV (1959) in their analysis of the influence of elastic
strains on the process of Ostwald ripening. The cited authors came to
the conclusion, that elastic strains only modify the kinetics of growth

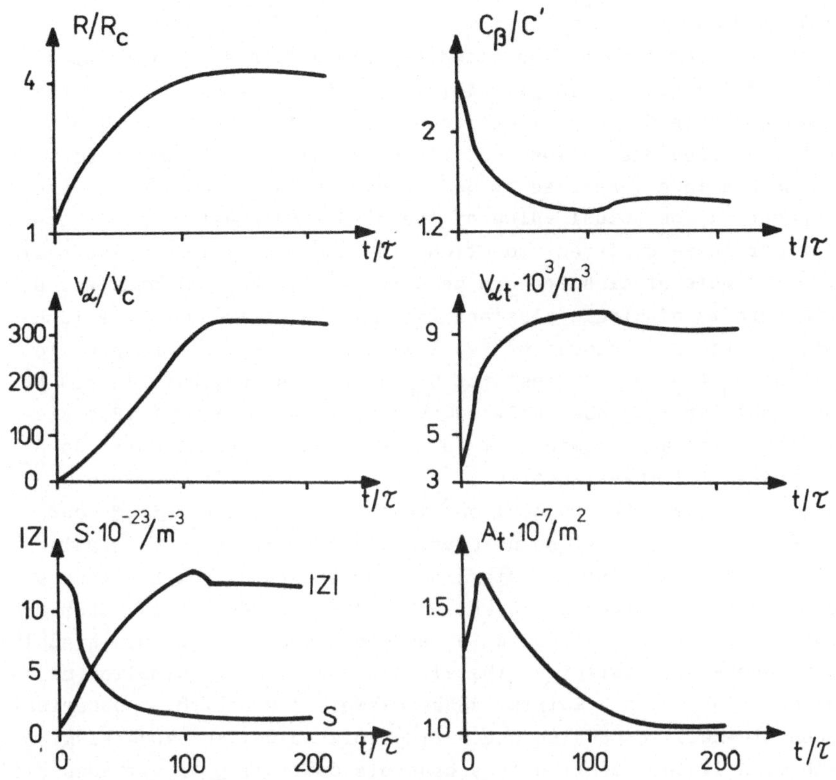

Fig. 8.7:
Ostwald ripening in elastic media. While in the first stage of the
ripening process ($V_\alpha \leqslant V_0$) elastic strains of the Nabarro-type (8.20)
only modify the evolution quantitatively, elastic strains of the form
(8.25) lead to a quite different asymptotic behaviour as compared with
the results shown in Fig. 6.1 ($\varkappa = 4.8 \cdot 10^8$ N/m^2, $\varepsilon = 2.37 \cdot 10^7$ N/m^2
for the values of the other parameters see Fig. 6.1.

quantitatively. We could show, that there exists, however, cases,
where elastic strains lead to a qualitatively different asymptotic
behaviour as compared with the results obtained by Lifshitz and Slyozov. That this is not only a theoretical possibility, but has practical relevance is demonstrated, e.g., by the experimental results of
PASCOVA and GUTZOW (1983). A comparison between Figs. 8.1 and 8.7 pro-
ves, that the theoretical and experimental curves are qualitatively
the same. Moreover, it can be shown, that for a special choice of the

parameter V_0 also a quantitative coincidence can be reached(SCHMELZER, GUTZOW, 1986a).

For an interpretation of the cited experimental results viscous properties of the matrix, in principle, have to be taken into account. This is possible by an application of eq. (8.39). The evolution of the ensemble of clusters is then described by the eqs. (8.50), again, where $\varkappa$ has to be replaced by $\varkappa''$. Since $\varkappa''$ is a function of time, in dependence on the actual value of the diffusion coefficient of the matrix particles different solutions for the mean value of the radius of the ensemble of clusters can be obtained. In an analogous way as for the growth of single clusters (see Fig. 8.6) the possible functions $R = R(t)$ are bounded by two limiting curves corresponding to $\varkappa = 0$ or $D_M \approx D$ and $\varkappa = const$ or $D_M \longrightarrow 0$, respectively. It can be shown, that for the case analyzed by PASCOVA and GUTZOW (1983) viscous properties are unimportant for the chosen times of duration of the experimental investigations.

Finally, we would like to mention, that the discussed methods can be applied to an investigation of crystallization processes of small particles, crystallization, starting from inner and outer surfaces. First results are given by HOELL, SCHMELZER, MAHNKE (1986), SCHMELZER (1985a). The authors calculate for an one component system, using the linear theory of elasticity, the elastic energy of deformation of a spherical cluster in a matrix. As an example the growth of spherical silicon clusters is considered. During this so-called Czochralski growth of silicon, occuring in a crucible consisting of vitreous silica, a formation of β-cristobalite (the high temperature modification of cristobalite) takes place in the volume around pores and at the surface. This effect is annouying for the growth of single crystals, because the strains connected with this effect lead to cracks and to the break away of particles in the glass. This reduces after all because of soilings and displacements the quality of the silicon single crystal.

Among the different boundary conditions which are considered the authors investigate in detail the influence of an additional hydrostatic pressure on the surface, $\sigma_{rr}(R) = -p$, and a surface tension between the interface cluster matrix. Results are shown in Fig. 8.5, for the explicit calculations see HOELL (1986).

References

ABRAHAM, F.F.: Homogeneous Nucleation Theory, New York, 1974

ABRAHAM, F.F.: Computer Simulations of microclusters, Surfaces, and Interfaces. In: Critical Review in Solid State and Material Sciences 1981, p. 169

ABRAHAM, F.F.: Statistical Surface Physics. A Perspective via Computer Simulation of Microclusters, Interfaces and Simple Films, Rep.Progr. Phys. 45 (1982) 1113

ABRAHAM, F.F., POUND, G.M.: Re-Examination of Homogeneous Nucleation Theory: Statistical Thermodynamic Aspects, J.Chem.Phys. 48 (1968) 732

BAIDAKOV, W.G., SKRIPOV, W.P.: Zh.Fiz.Khim. 56 (1982) 1234

BAKKER, G.: Kapillarität und Oberflächenspannung, Handbuch der Experimentalphysik, Bd. 6, Hrsg.: W. Wien, F. Harms, H. Lenz, Leipzig, 1928

BARSCHDORFF, D.: Carrier Gas Effects on Homogeneous Nucleation of Water Vapor in a Shock Tube, Phys.Fluids 18 (1975) 529

BECKER, C., REISS, H.: Nucleation in a Nonuniform Vapor, J.Chem.Phys. 65 (1976) 2066

BECKER, R.: Theorie der Wärme, Berlin, Göttingen, Heidelberg, 1961

BECKER, R., DÖRING, W.: Kinetische Behandlung der Keimbildung in übersättigten Dämpfen, Ann.Physik 24 (1935) 719

BEENAKKER, C.W.J., ROSS, J.: Theory of Ostwald Ripening for Open Systems, J.Chem.Phys. 83 (1985) 4710

BELOSLUDOW, W.R., NABUTOWSKIJ, W.P.: Zh.Exper.Teor.Fiz. 68 (1975) 2177

BHARUCHA-REID, A.T.: Elementy Markovskich Prozessov i ich Priloshenija, Moscow, 1969

BINDER, K.: Theory for the Dynamics of Clusters, II. Critical Diffusion in Binary Liquids and the Kinetics of Phase Separation, Phys.Rev. B25 (1977) 4425 (a)

BINDER, K.: Computer Experiments on Nucleation Processes in the Lattice Gas Model, Adv.Coll.Interface Sci. 7 (1977) 279 (b)

BINDER, K. (Ed.): Monte-Carlo Methods in Statistical Physics, Berlin (W.), Heidelberg, New York, Vol. 1, 1979, Vol. 2, 1984

BINDER, K., BILLOTET, C., MIROLD, P.: On the Theory of Spinodal Decomposition in Solid and Liquid Binary Mixtures, Z.Physik B30 (1978) 183

BINDER, K., STAUFFER, D.: Statistical Theory of Nucleation, Condensation and Coagulation, Adv.Phys. 25 (1976) 343

BLANDER, M., HENGSTENBERG, D., KATZ, J.L.: Bubble Nucleation in n-Pentane, n-Hexane, n-Pentane/Hexadecane Mixtures, and Water, J.Chem.Phys. 75 (1971) 3631

BUDDE, A., MAHNKE, R.: Isoenergetic Nucleation in Finite Systems, Rostocker Phys. Manuskr. 10 (1987) 92

BUDDE, A., PASCOVA, R., MAHNKE, R., GUTZOW, I.: Modelluntersuchungen des Wachstums von Silberchloridausscheidungen in fotochromen Gläsern, Wiss.Z.WPU Rostock, NR 35 (1986) H.4, 24

BUFF, F.P.: The Spherical Interface, I. Thermodynamics, J.Chem.Phys. 19 (1951) 1591

CAHN, J.W.: On Spinodal Decomposition, Acta Met. 9 (1961) 795

CAHN, J.W.: On Spinodal Decomposition in Cubic Crystals, Acta Met. 10 (1962) 179

CAHN, J.W.: Phase Separation by Spinodal Decomposition in Isotropic Systems, J.Chem.Phys. 42 (1965) 93

CAHN, J.W.: The Later Stages of Spinodal Decomposition and the Beginning of Particle Coarsening, Acta Met. 14 (1966) 1685

CAHN, J.W.: Spinodal Decomposition, Trans.Metall.Soc. AIME 242 (1968) 166

CAHN, J.W.: Critical Point Wetting, J.Chem.Phys. 66 (1977) 3667

CAHN, J.W., HILLIARD, J.E.: Free Energy of a Nonuniform System, I. Interfacial Free Energy, III. Nucleation in a Two-Component Incompressible Liquid, J.Chem.Phys. 28 (1958) 258; 31 (1959) 688

CASTLEMAN, A.W., KEESEE, R.G.: Clusters: Bridging the Gas and Condensed Phases, Acc.Chem.Res. 19 (1986) 413

CHAKRAVERTY, B.K.: Grain Size Distribution in Thin Films, I. Conservation Systems, II. Non-Conservative Systems, J.Phys.Chem.Sol. 28 (1967) 2401, 2413

CHRISTIAN, J.W.: The Theory of Transformations in Metals and Alloys, (2nd Edition), Oxford, 1975

COOK, H.E.: Brownian Motion in Spinodal Decomposition, Acta Met. 18 (1970) 297

DEFAY, R., PRIGOGINE, I., BELLEMANS, R., EVERETT, D.H.: Surface Tension and Adsorption, London, 1966

DERJAGUIN, B.V., PROKHOROV, A.V.: Improved Theory of Homogeneous Condensation and its Comparison with Experimental Data, J.Colloid Interface Sci. 46 (1974) 283

DOMB, C., GREEN, M.S., LEBOWITZ, J.L.: Phase Transitions and Critical Phenomena, Vol.1-8, London, New York, 1972-1983

EBELING, W.: Thermodynamisch-statistische Fluktuationen und Kinetik der Keimbildung, Sitzungsberichte der AdW der DDR, 27N (1981) 33

EBELING, W., BUDDE, A., MAHNKE, R., SCHIMANSKY-GEIER, L.: On the Kinetics of Nucleation in Isolated Gaseous Systems, in preparation, 1988

EBELING, W., ENGEL, A., ESSER, B., FEISTEL, R.: Diffusion and Reaction in Random Media and Models of Evolution Processes, J.Stat.Phys. 37 (1984) 369

EBELING, W., FEISTEL, R.: Physik der Selbstorganisation und Evolution, Berlin, 1982

EBELING, W., IVANOV, C., SCHIMANSKY-GEIER, L.: Stochastic Theory of Nucleation in Bistable Reaction Systems, Rostocker Phys.Manuskr. 2 (1977) 93

EBELING, W., KLIMONTOVICH, Y.L.: Selforganization and Turbulence in Liquids, Teubner-Texte zur Physik, Bd.2, Leipzig, 1984

EBELING, W., KREMP, D., KRAEFT, W.D.: Theory of Bound States, Berlin, 1976, Russian Edition, Moscow, 1979

EBELING, W., SCHIMANSKY-GEIER, L.: Stochastic Dynamics of a Bistable Reaction System, Physica 98A (1979) 587

EBELING, W., SCHIMANSKY-GEIER, L.: Nonequilibrium Phase Transitions and Nucleation in Reacting Systems, Proc. 6th Int.Conf.on Thermodynamics, Merseburg, 1980, p. 95

EBELING, W., SCHMELZER, J.: Koexistenz von Sorten in nichtlinearen autokatalytischen Reaktionssystemen, Z.Phys.Chemie (Leipzig) 261 (1980) 677 (a)

EBELING, W., SCHMELZER, J.: Contributions to the Theory of Competing Predators I, studia biophysica 78 (1980) H.1, 31 (b)

EBELING, W., ULBRICHT, H. (Eds.): Selforganization by Nonlinear Irreversible Processes, Springer Series in Synergetics, Vol.33, Berlin (W) 1986

EHRENFEST, P.: Comm. Leiden Univ. 20 (1933) Suppl. b 75, 628

EIGEN, M.: Selforganization of Matter and the Evolution of Biological Macromolecules, Naturwiss. 58 (1971) 465

EIGEN, M.; SCHUSTER, P.: The Hypercycle, Berlin (W.), 1979

FALLS, A.H., SCRIVEN, L.E., DAVIS, H.T.: Structure and Stress in Spherical Microstructures, J.Chem.Phys. 75 (1981) 3986

FALLS, A.H., SCRIVEN, L.E., DAVIS, H.T.: Adsorption, Structure and Stress in Binary Interfaces, J.Chem.Phys. 78 (1983) 7300

Faraday Symposia of the Chemical Society: Structure of the Interfacial Region, Faraday Symp.Chem.Soc. 16 (1981)

FEINN, D., ORTOLEVA, P., SCALF, W., SCHMIDT, S., WOLFF, M.: Spontaneous Pattern Formation in Precipitating Systems, J.Chem.Phys. 69 (1978) 27

FEISTEL, R.: Selektion und nichtlineare Oszillationen in chemischen Modellreaktionen, Diss.B, Rostock, 1979

FEISTEL, R.: Nonlinear Chemical Reactions in Diluted Solutions. In: Selforganization by Nonlinear Irreversible Processes (Eds.: W. Ebeling, H. Ulbricht), Berlin (W.), 1986, p. 48

FEISTEL, R., EBELING, W.: Deterministic and Stochastic Theory of Sustained Oscillations in Autocatalytic Reaction Systems, Physica 93A (1978) 114

FEISTEL, R., EBELING, W.: Dynamische Modelle zum Selektionsverhalten offener Systeme, Wiss.Z.WPU Rostock, MNR 25 (1971) H.5, 507

FICHTENHOLZ, G.M.: Differential- und Integralrechnung, Berlin, 1966; Russian Edition, Moscow, 1959

FISHER, L.R., ISRAELACHVILI, J.N.: Determination of the Capillarity Pressure in Menisci of Molecular Dimensions, Chem.Phys.Lett. 76 (1980) 325

FLEMING, G.D., COURTNEY, S.H., BALK, M.W.: Activated Barrier Crossing: Comparison of Experiment and Theory, J.Stat.Phys. 42 (1986) 83

FRENKEL, J.: Kinetic Theory of Liquids, New York, 1955; Edition in Russian, Leningrad, 1975

GARDINER, C.W.: Handbook of Stochastic Methods, Berlin (W.), Heidelberg, New York, 1984

GEBHARDT, W., KREY, U.: Phasenübergänge und kritische Phänomene, Braunschweig, 1980

GEBRE MEDHIN, T., SCHMELZER, J.: On the Kinetics of First-Order Phase Transitions in Adsorbed Layers, to be published, 1987

GIBBS, J.W.: On the Equilibrium of Heterogeneous Substances, Trans. Conn.Acad.Sci. 3 (1875-78) 108, 343; see also: J.W. Gibbs, Collected Works, vol.1, Thermodynamics, New York-London-Toronto, 1928, Russian Translation, Moskau, 1982

GILLESPIE, D.T.: A Pedestrian Approach to Transitions and Fluctuations in Simple Nonequilibrium Chemical Systems, Physica 95A (1979) 69

GILLESPIE, D.T.: Transition Time Statistics in Simple Bistable Chemical Systems, Physica 101A (1980) 535

GILLESPIE, D.T.: A Stochastic Analysis of the Homogeneous Nucleation of Vapor Condensation, J.Chem.Phys. 74 (1981) 661

GOLDBURG, W.I., HUANG, J.S.: Phase Separation Experiments Near the Critical Point, In: Fluctuations, Instabilities and Phase Transitions (Ed.: T. Riste), New York-London, 1975

GOODRICH, F.C., RUSANOV, A.I. (Eds.): The Modern Theory of Capillarity, Berlin, 1981

GRAHAM, R.: Springer Tracts in Modern Physics 66 (1973) 1

GREENWOOD, G.W.: The Growth of Dispersed Pricipitates in Solutions, Acta Met. 4 (1956) 243

GUERMEUR, R., BIQUARD, F., JACOLIN, C.: Density Profiles and Surface Tension of Spherical Interfaces: Numerical Results for Nitrogen Drops and Bubbles, J.Chem.Phys. 82 (1985) 2040

GUGGENHEIM, E.A.: The Thermodynamics of Interfaces in Systems of Several Components, Trans.Faraday Soc. 36 (1940) 397; 41 (1945) 150; E.A. Guggenheim, Thermodynamics, Amsterdam, 1949

GUNTON, J.D., SAN MIGUEL, M., SAHNI, P.S.: The Dynamics of First-Order Phase Transitions. In: Phase Transitions, Vol.8, (Eds.: C. Domb, J.L. Lebowitz), 1983, p. 267

GUROV, K.P.: Phenomenological Thermodynamics of Irreversible Processes, Moscow, 1978 (in Russian)

GUTZOW, I., AVRAMOV, I.: The Mechanism of Formation, the Structure and the Properties of Amorphous Films, Thin Solid Films 85 (1981) 203

GUTZOW, I., SCHMELZER, J., PASCOVA, R., POPOW, B.: Nucleation and Crystal Growth in Viscoelastic Media, Rostocker Phys.Manuskr. 8 (1985) 22

HAASE, R.: Thermodynamik der irreversiblen Prozesse, Darmstadt, 1963

HAKEN, H.: Synergetics. An Introduction, Berlin (W.), Heidelberg, New York, 1978

HALE, B.H., WARD, R.C.: A Monte-Carlo Method for Approximating Critical Cluster Size in the Nucleation of Model Systems, J.Stat.Phys. 28 (1982) 487

HÄNGGI, P.: Escape from a Metastable State, J.Stat.Phys. 42 (1986) 105

HANSZEN, K.-J.: Theoretische Untersuchungen über den Schmelzpunkt kleiner Kügelchen, Z.Physik 157 (1960) 523

HEADY, R.B., CAHN, J.W.: Experimental Test of Classical Nucleation Theory in a Liquid-Liquid Miscibility Gap System, J.Chem.Phys. 58 (1973) 896

HEERMANN, D.W., CONIGLIO, A., KLEIN, W., STAUFFER, D.: Nucleation and Metastability in Three-Dimensional Ising Models, J.Stat.Phys. 36 (1984) 447

HEINRICH, M., ULBRICHT, H.: Mechanik der Kontinua (WTB, Bd. 182), Berlin, 1981

HEIST, R.M., COOLING, K.M., DUPUIS, C.S.: Homogeneous Nucleation in Associated Vapours, I. Acetic Acid, J.Chem.Phys. 65 (1976) 5147

HEMINGWAY, S.J., HENDERSON, J.R., ROWLINSON, J.S.: The Density Profile and Surface Tension of a Liquid Drop, Faraday Symposia Chem.Soc. 16 (1981) 33

HENDERSON, J.R.: Statistical Mechanics of Spherical Surfaces, In:
Fluid Interfacial Phenomena (Ed.: C.A. Croxton), New York, 1986,
p. 555

HENDERSON, J.R., ROWLINSON, J.S.: Statistical Mechanics of Fluid
Interfaces in Cylindrical Symmetry, J.Phys.Chem. 88 (1984) 6484

HEYES, D.M., WOODCOCK, L.V.: Clustering and Some Other Physical Effects of Van der Waals Potentials, Molec.Phys. 59 (1986) 1369

HILL, T.L.: Thermodynamics of Small Systems, New York-Amsterdam,
1963

HILLERT, M.: A Solid-Solution Model for Inhomogeneous Systems, Acta
Met. 9 (1961) 525

HIRTH, J.P., POUND, G.M.: Condensation and Evaporation, Nucleation and
Growth Kinetics, Vol. 11, Progress in Material Science (Ed.: B.
Chalmers), London, 1963

HODGSON, A.W.: Homogeneous Nucleation, Adv.Colloid Interface Sci.,
21 (1984) 303

HOELL, A.: Einfluß elastischer Spannungen auf das Keimwachstum,
Diplomarbeit, Wilhelm-Pieck-Universität Rostock, 1986

HOELL, A., SCHMELZER, J., MAHNKE, R.: On the Influence of Elastic
Strains on the Crystallization in Glasses, Wiss.Z.WPU Rostock, NR
35 (1986) H.4, 28

HOHMANN, H.H.: Untersuchungen zur Ostwaldreifung, Diss., Göttingen,1974

HYNES, J.T.: Chemical Reaction Rates and Solvent Friction, J.Stat.Phys.
42 (1986) 149

JACKSON, K.A.: Crystal Growth Kinetics, Material Science and Engineering 65 (1984) 7

JAIN, S.C., HUGHES, A.E.: Ostwald Ripening and its Applications to
Precipitates and Colloids in Ionic Crystals and Glasses, J.Material
Sci. 13 (1978) 1611

JANSSEN, H.K.: Stochastisches Reaktionsmodell für einen Nichtgleich-
gewichts-Phasenübergang, Z.Physik 270 (1974) 67

JORTNER, J.: Level Structure and Dynamics of Clusters, Ber.Bunsenges.
Phys.Chem. 88 (1984) 188

KAGAN, Ju.: Die Kinetik des Siedens einer reinen Flüssigkeit,
Zh.Fiz.Khim. 34 (1960) 92

KAHLWEIT, M.: Precipitation and Aging. In: Physical Chemistry, Vol.10,
New York, 1970, p. 719

KAHLWEIT, M.: Ostwald Ripening of Precipitates, Adv.Colloid Interface
Sci. 5 (1975) 1

KALOS, M., LEBOWITZ, J.L., PENROSE, O., SUR, A.: Clusters, Metastability, and Nucleation: Kinetics of First-Order Phase Transitions, J.Stat.
Phys. 18 (1978) 39

KASHCHIEV, D.: Solution of the Non-Steady State Problem in Nucleation
Theory, Surface Science 14 (1969) 209

KATZ, J.L.: Condensation of a Supersaturated Vapor. I. The Homogeneous
Nucleation of the n-Alcanes, J.Chem.Phys. 52 (1970) 4733

KATZ, J.L., BLANDER, M.: Condensation and Boiling: Corrections to Homogeneous Nucleation Theory for Non-Ideal Gases, J.Colloid Interface
Sci. 42 (1973) 496

KATZ, J.L., DONOHUE, M.D.: A Kinetic Approach to Homogeneous Nucleation
Theory, Adv.Chem.Phys. 40 (1979) 137

KATZ, J.L., DONOHUE, M.D.: Nucleation with Simulataneous Chemical
Reactions, J.Colloid Interface Sci. $\underline{85}$ (1982) 267

KATZ, J.L., MIRABEL, P., SCOPPA, C.J., VIRKLER, T.L.: Condensation of
a Supersaturated Vapour. III. The Homogeneous Nucleation of CCl_4,
$CHCl_3$, CCl_3F and $O_2H_2Cl_4$, J.Chem.Phys. $\underline{65}$ (1976) 382

KATZ, J.L., OSTERMIER, B.J.: Diffusion Cloud-Chamber Investigation of
Homogeneous Nucleation, J.Chem.Phys. $\underline{47}$ (1967) 478

KATZ, J.L., SCOPPA, c.J., KUMAR, H.G., MIRABEL, P.: Condensation of a
Supersaturated Vapor. II. The Homogeneous Nucleation of the n-Alkyl
Benzenes, J.Chem.Phys. $\underline{62}$ (1975) 448

KATZ, J.L., WIEDERSICH, H.: J.Colloid Interface Sci. $\underline{61}$ (1977) 351

KAWASAKI, K., ENOMOTO, Y., TOKUYAMA, M.: Elementary Derivation of Kinetic Equations for Ostwald Ripening, Physica $\underline{135}$ A (1986) 426

KEENEY, M., HEICKLEN, J.: Surface Tension and the Heat of Vaporization.
A Simple Empirical Correlation, J.Inorg.Nucl.Chem. $\underline{41}$ (1979) 1755

KELTON, K.F., GREER, A.L., THOMPSON, C.V.: Transient Nucleation in
Condensed Systems, J.Chem.Phys. $\underline{79}$ (1983) 6281

KERLEY, G.I.: Theory of Ionization Equilibrium: An Approximation for
the Single Element Case, J.Chem.Phys. $\underline{85}$ (1986) 5228

KIRKWOOD, J.G., BUFF, F.P.: The Statistical Mechanical Theory of Surface Tension, J.Chem.Phys. $\underline{17}$ (1949) 338

KNESSL, C., MANGEL, M., MATKOWSKY, B.J., SCHUSS, Z., TIER, C.:
J.Chem.Phys. $\underline{81}$ (1984) 1285

KNESSL, C., MATKOWSKY, B.J., SCHUSS, Z., TIER, C.: A Singular Pertur-
bation Approach to First Passage Times for Markov Jump Processes,
J.Stat.Phys. $\underline{42}$ (1986) 169

KOBRAEI, H.R.: A Model of Interacting Clusters for Microscopic Nucleation Theory, J.Chem.Phys. $\underline{81}$ (1984) 6051

KOCH, S.W., DESAI, R.C., ABRAHAM, F.F.: Kinetics of Nucleation in
Near-Critical Fluids, Phys.Rev. A $\underline{27}$ (1983) 2152

KOCH, S.W., LIEBMANN, R.: Comparison of Molecular Dynamics and Monte-
Carlo Computer Simulations of Spinodal Decompositions, J.Stat.Phys.
$\underline{33}$ (1983) 31

KÖNOBEJEWSKI, S.: Zur Theorie der unterkühlten festen Lösungen,
Z.Phys.Chem. $\underline{171}$ (1934) 25

KOTAKE, S., GLASS, I.I.: Survey of Flows with Nucleation and Condensation, UTIAS Rev. Nr. 42, 1978

KRAMERS, A.: Physica (Utrecht) $\underline{7}$ (1940) 284

KRISHNAMURTHY, S., GOLDBURG, W.I.: Kinetics of Nucleation in a Binary
Liquid Mixture, Phys.Rev. A $\underline{22}$ (1980) 2147

KUBO, R.: Thermodynamics, Amsterdam, 1968

LAUMANN, B.: Z.Naturforschung $\underline{17a}$ (1962) 808, 812

LANDAU, L.D., LIFSHITZ, E.M.: Lehrbuch der theoretischen Physik,
Bd.VII. Elastizitätstheorie, Berlin, 1977

LANDAU, L.D., LIFSHITZ, E.M.: Lehrbuch der theoretischen Physik,
Bd.V. Statistische Physik (5., bearbeitete Auflage), Berlin, 1979

LANDAU, L.D., LIFSHITZ, E.M.: Lehrbuch der theoretischen Physik, Bd. X.
Physikalische Kinetik, Berlin, 1983

LANGER, J.S.: Theory of Condensation Point, Ann.Physics $\underline{41}$ (1967)
108

LANGER, J.S.: Statistical Theory of Decay of Metastable States,
Ann.Physics 54 (1969) 258

LANGER, J.S.: Kinetics of Metastable States, Lecture Notes in Physics,
132 (1980) 12

LANGER, J.S., BAR-ON, M., MILLER, H.L.: New Computational Method in
the Theory of Spinodal Decomposition, Phys.Rev. A 11 (1975) 1417

LANGER, J.S., FISHER, M.E.: Intrinsic Critical Velocity of a Super-
fluid, Phys.Rev.Lett. 19 (1967) 560

LANGER, J.S., SCHWARTZ, A.J.: Kinetics of Nucleation in Near-Critical
Fluids, Phys.Rev. A 21 (1980) 948

LANGER, J.S., TURSKI, L.A.: Hydrodynamic Model of the Condensation of
a Vapor Near its Critical Point, Phys.Rev. A 8 (1973) 3230

LIESEGANG, R.E.: Z.Phys.Chemie 75 (1911) 374

LIFSHITZ, I.M., SLYOZOV, V.V.: On the Kinetics of Diffusional Decay
of Supersaturated Solid Solutions, Zh.Exper.Teor.Fiz. 35 (1958) 479

LIFSHITZ, I.M., SLYOZOV, V.V.: Zur Theorie der Koaleszens fester
Lösungen, Fiz.Tverdovo Tela 1 (1959) 1401

LIFSHITZ, I.M., SLYOZOV, V.V.: The Kinetics of Precipitation from
Supersaturated Solid Solutions, J.Phys.Chem.Solids 19 (1961) 35

LINDENBERG , K., SESHADRI, V.: Analytic Theory of Extrema: I. Asymptotic Theory for Fokker-Planck Processes, J.Chem.Phys. 71 (1979) 4075

LOTHE, J., POUND, G.M.: On the Statistical Mechanics of Nucleation
Theory, J.Chem.Phys. 45 (1966) 630

LOVETT, R.: The Rate of Nucleation in a Vapor-Liquid Transition Determined from Hydrodynamic Fluctuation Theory. J.Chem.Phys. 81 (1984) 6191

LOVETT, R., ORTOLEVA, P., ROSS, J.: Kinetic Instabilities in First-
Order Phase Transitions, J.Chem.Phys. 69 (1978) 947

LUDIG, A.: Untersuchung zur spontanen Kondensation bei stationärer
Überschallströmung unter Berücksichtigung des Realgasverhaltens, Diss.
Universität Karlsruhe, 1975

MAHNKE, R.: Zur Komplexität und Evolution biologischer Makromoleküle,
Diss.A, Wilhelm-Pieck-Universität Rostock, 1979

MAHNKE, R.: Dynamics of First-Order Phase Transitions in Finite Systems, Wiss.Z.WPU Rostock, NR 35 (1986) H.4, 13

MAHNKE, R.: On the Kinetics of Ostwald Ripening in Finite Systems,
Rostocker Phys.Manuskr. 10 (1987)

MAHNKE, R., BUDDE, A.: Keimbildung in isothermen Systemen, Wiss.Z.
WPU Rostock, NR 36 (1987) (im Druck)

MAHNKE, R., BUDDE, A.: Zur Kinetik der Keimbildung bei isotherm-iso-
choren Randbedingungen, Wiss.Z.WPU Rostock, in press, 1988

MAHNKE, R., FEISTEL, R.: The Kinetics of Ostwald Ripening as a Competitive Growth in a Selforganizing System, Rostocker Phys.Manuskr.
8 (1985) 54

MAHNKE, R., SCHMELZER, J.: A New General Formula for the Curvature
Dependence of Surface Tension of Droplets, Z.Phys.Chemie (Leipzig)
266 (1985) 1028

MAHNKE, R., SCHMELZER, J., THIESSENHUSEN, K.-U.: Oberflächenspannung
kleiner Teilchen, Wiss.Z.WPU Rostock, NR 34 (1985), H.1, 48

MALCHOW, H., SCHIMANSKY-GEIER, L.: Noise and Diffusion in Bistable
Nonequilibrium Systems, Teubner-Texte zur Physik, Bd.5, Leipzig,
1985

MARQUSEE, J.A.: Dynamics of Late Stage Phase Separation in Two Dimensions, J.Chem.Phys. $\underline{18}$ (1984) 976

MARQUSEE, J.A., ROSS, J.: Kinetics of Phase Transitions. Theory of Ostwald Ripening, J.Chem.Phys. $\underline{79}$ (1983) 373

MARQUSEE, J.A., ROSS, J.: Theory of Ostwald Ripening. Competitive Growth and its Dependence on Volume Fraction, J.Chem.Phys. $\underline{80}$ (1984) 536

MATHESON, J., WALLS, D.F., GARDINER, C.W.: J.Stat.Phys. $\underline{19}$ (1978) 65

MATKOWSKY, B.J., SCHUSS, Z., KNESSL, C., TIER, C., MANGEL, M.: Asymptotic Solution of the Kramers-Moyal Equation and First-Passage Times for Markov Jump Processes, Phys.Rev. $\underline{29A}$ (1984) 3359

MATZ, G.: Kristallisation, Springer, 1969

MILCHEV, A., BINDER, K., HEERMANN, D.W.: Fluctuations and Lack of Self-Averaging in the Kinetics of Domain Growth, Z.Physik B $\underline{63}$ (1986) 521

MILCHEV, A., MARKOV, I.: The Effect of Anharmonicity in Epitaxial Interfaces, Surface Sci. $\underline{136}$ (1984) 503, 519

MIROLD, P., BINDER, K.: Theory for the Initial Stages of Grain Growth and Unmixing Kinetics of Binary Alloys, Acta Met. $\underline{25}$ (1977) 1435

MONTROLL, E.W., SHULER, K.E.: Adv.Chem.Phys. $\underline{1}$ (1958) 361

MUTAFTSCHIEV, B.: Surface Thermodynamics. In: Interfacial Aspects of Phase Transformations, Proc.Nato Advanced Study Instituts, Dordrecht, Netherlands, 1982

NABARRO, F.R.N.: The Influence of Elastic Strain on the Shape of Particle Segregating in an Alloy, Proc.Phys.Soc. $\underline{52}$ (1940) 90; Proc.Roy. Soc. A $\underline{175}$ (1940) 519

NEUMANN, K., DÖRING, W.: Tröpfchenbildung in übersättigten Dampfgemischen zweier vollständig mischbarer Flüssigkeiten, Z.Phys.Chem. $\underline{186}$ A (1940) 203

NICOLIS, G., PRIGOGINE, I.: Selforganization in Non-Equilibrium Systems, New York, 1977

NISHIOKA, K., POUND, G.M.: Statistical Mechanics of Homogeneous Nucleation in Vapor. In: Nucleation Phenomena (Ed.: A.C. Zettlemoyer), 1977

NONNENMACHER, T.F.: Size Effects on Surface Tension and Vapour Pressure of Small Droplets, University of Ulm Report Nr. 12.1, 1976

NONNENMACHER, T.F.: Size Effect on Surface Tension of Small Droplets, Chem.Phys.Lett. $\underline{47}$ (1977) 507

ONO, S., KONDO, S.: Molecular Theory of Surface Tension in Liquids, In: Handbuch der Physik, Bd.10, Hrsg.: S. Flügge, Berlin (W.), 1960

ONSAGER, L.: Phys.Rev. $\underline{65}$ (1944) 117

OPPENHEIM, I., SHULER, K.E., WEISS, G.H.: Stochastic Theory of Nonlinear Rate Processes with Multiple Stationary States, Physica $\underline{88A}$ (1977) 191

OSTWALD, W.: Analytische Chemie, Leipzig, 1901

PARLANGE, J.-Y.: Free Energy of Formation of Droplets with Curvature Dependent Surface Tension, J.Crystal Growth $\underline{6}$ (1970) 311

PASCOVA, R., GUTZOW, I.: Modelluntersuchungen zum Bildungsmechanismus fototroper Silberhalogenidphasen in Gläsern, Glastechnische Berichte $\underline{56}$ (1983) 324

PATASHINSKY, A.Z., POKROVSKI, V.L.: Fluktuačionnaja Theorija Fazovych
Perechodov, Moskva, 1982 (in Russian)

PATASCHINSKY, A.Z., SHUMILO, B.I.: Theorie der Relaxation metastabiler
Zustände, Zh.Exper.Teor.Fiz. $\underline{77}$ (1979) 1417

PATASCHINSKY, A.Z., SHUMILO, B.I.: Relaxation metastabiler Zustände
in der Umgebung eines trikritischen Punktes, Fiz.Tverdovo Tela $\underline{22}$
(1980) 2609

PENROSE, O., BUHAGIAR, A.: Kinetics of Nucleation in a Lattice Gas
Model. Microscopic Theory and Simulation Compared, J.Stat.Phys. $\underline{30}$
(1983) 219

PENROSE, O., LEBOWITZ, J.L., MARRO, J., KALOS, M.H., SUR, A.:
Growth of Clusters in a First-Order Phase Transition, J.Stat.Phys.
$\underline{19}$ (1978) 243

PENROSE, O.; LEBOWITZ, J.L., MARRO, J., KALOS, M., TOBOCHNIK, J.:
Kinetics of a First-Order Phase Transition. Computer Simulation and
Theory, J.Stat.Phys. $\underline{34}$ (1984) 399

PHILLIPS, P.: On the Surface Tension of Liquid Drops, Molec.Phys. $\underline{52}$
(1984) 805

PRIGOGINE, I., BELLEMANS, A.: Statistical Mechanics of Surface Tension
and Adsorption. In: Adhesion and Adsorption of Polymers, Part A
(Ed.: L.-H. Lee), New York, 1980

PROCACIA, I., ROSS, J.: J.Chem.Phys. $\underline{67}$ (1977) 5565

RAO, M., BERNE, B.J., KALOS, M.H.: Computer Simulation of the Nucleation and Thermodynamics of Microclusters, J.Chem.Phys. $\underline{68}$ (1978) 1325

RASMUSSEN, D.H.: Energetics of Homogeneous Nucleation, J.Crystal
Growth $\underline{56}$ (1982) 45

RASMUSSEN, D.H.: Thermodynamics and Nucleation Phenomena, J.Crystal
Growth $\underline{56}$ (1982) 56

RASMUSSEN, D.H., APPELBY, M.R., LEEDOM, G.L., BABU, S.V., NAUMANN,
R.J.: Homogeneous Nucleation Kinetics, J.Crystal Growth $\underline{64}$ (1983)
229

RAWSON, H.: Inorganic Glass-Forming Systems, London-New York, 1967

REISS, H., KATZ, J.L.: Resolution of the Translation-Rotation Paradox
in the Theory of Irreversible Condensation, J.Chem.Phys. $\underline{46}$ (1967) 2496

REISS, H., KATZ, J.L., COHEN, E.R.: Translation-Rotation Paradox in the
Theory of Nucleation, J.Chem.Phys. $\underline{48}$ (1968) 5553

ROLOV, B.N., JURKEWITSCH, W.E.: Physik der diffusen Phasenübergänge,
Rostow, 1983 (in Russian)

RÖPKE, G.: Statistische Mechanik für das Nichtgleichgewicht, Berlin, 1987

ROWLINSON, J.S., WIDOM, B.: Molecular Theory of Capillarity, Oxford, 1982

RUSANOV, A.I.: Phasengleichgewichte und Grenzflächenerscheinungen,
Leningrad, 1967 (in Russian); Extended Edition in German, Berlin, 1978

RUSANOV, A.I.: Surface and Cohesive Energy of Condensed Bodies,
J.Colloid Interface Sci. $\underline{90}$ (1982) 147

RUSANOV, A.I., BRODSKAYA, E.N.: The Molecular Dynamics Simulation of
a Small Drop, J.Colloid Interface Sci. $\underline{62}$ (1977) 543

SANCHEZ, I.C.: Liquids: Surface Tension, Compressibility and Invariants, J.Chem.Phys. $\underline{79}$ (1983) 405

SAN MIGUEL, M., GRANT, M., GUNTON, J.D.: Phase Separation in Two-Dimensional Binary Fluids, Phys.Rev. A $\underline{31}$ (1985) 1001

SCHERBAKOW, L.M., RJASANZEW, P.P., FILIPPOWA, N.P.: Kolloidnyi Zhurnal $\underline{23}$ (1961) 338

SCHIMANSKY-GEIER, L., EBELING, W.: Stochastic Theory of Nucleation in Nonequilibrium Bistable Reaction Systems, Ann.Physik $\underline{40}$ (1983) 10

SCHIMANSKY-GEIER, L., SCHWEITZER, F., EBELING, W., ULBRICHT, H.: On the Kinetics of Nucleation in Isochoric Gases. In: Selforganization by Nonlinear Irreversible Processes (Eds.: W. Ebeling, H. Ulbricht), Berlin (W.), 1986, p. 67

SCHMELZER, J.: Zur Koexistenz von Sorten in der nichtlinearen Kinetik homogener Konkurrenzreaktionen, Diss.A, Wilhelm-Pieck-Universität Rostock, 1979

SCHMELZER, J.: Zur Anwendung der Gibbsschen Theorie der Oberflächeneffekte auf die Beschreibung von Phasenübergängen, Wiss.Z.WPU Rostock, NR 33 (1984) H.3, 40

SCHMELZER, J.: Thermodynamik finiter Systeme und die Kinetik von thermodynamischen Phasenübergängen 1. Art, Diss.B, Wilhelm-Pieck-Universität Rostock, 1985 (a)

SCHMELZER, J.: Curvature Dependent Surface Tension and the Work of Formation of Critical Cluster, Rostocker Phys.Manuskr. $\underline{8}$ (1985) 80 (b)

SCHMELZER, J.: Zur Thermodynamik heterogener Systeme, Wiss.Z.WPU Rostock, NR $\underline{34}$ (1985), H.1, 39 (c)

SCHMELZER, J.: Zur Kinetik des Keimwachstums in Lösungen, Z.Phys. Chemie (Leipzig) $\underline{266}$ (1985) 1957 (d)

SCHMELZER, J.: Zur Kinetik des Wachstums von Tropfen in der Gasphase, Z.Phys.Chemie (Leipzig) $\underline{266}$ (1985) 1121 (e)

SCHMELZER, J.: The Curvature Dependence of Surface Tension of Small Droplets, J.Chem.Soc., Faraday Trans. 1, $\underline{82}$ (1986) 1421 (a)

SCHMELZER, J.: Formation and Growth of Bubbles in One-Component Closed Isochoric Systems, Z.Phys.Chemie (Leipzig), subm. for publication, 1986 (b)

SCHMELZER, J., EBELING, W.: Contributions to the Theory of Competing Predators II, studia biophysica $\underline{80}$ (1980), H.1, 53

SCHMELZER, J., EBELING, W., FEISTEL, R.: Contributions to the Theory of Competing Predators III, studia biophysica $\underline{82}$ (1981) H.2, 157

SCHMELZER, J., GUTZOW, I.: Ostwald Ripening in Viscoelastic Media. In: Selforganization by Nonlinear Irreversible Processes (Eds.: W. Ebeling, H. Ulbricht), Berlin (W.), 1986), p. 144 (a)

SCHMELZER, J., GUTZOW, I.: On the Kinetic Description of Ostwald Ripening in Elastic Media, Wiss.Z.WPU Rostock, NR $\underline{35}$ (1986), H.4, 5 (b)

SCHMELZER, J., MAHNKE, R.: General Formulae for the Curvature Dependence of Droplets and Bubbles, J.Chem.Soc., Faraday Trans. 1, $\underline{82}$ (1986) 1413

SCHMELZER, J., SCHWEITZER, F.: Thermodynamik und Keimbildung. In: Wiss.Berichte des Zentralinstituts für Festkörperphysik und Werkstoffforschung der AdW der DDR $\underline{19}$ (1982) 144

SCHMELZER, J., SCHWEITZER, F.: Thermodynamik und Keimbildung. I. Isotherme Keimbildung in finiten Systemen, Z.Phys.Chemie (Leipzig) $\underline{266}$ (1985) 943

SCHMELZER, J., SCHWEITZER, F.: Ostwald Ripening of Bubbles in
Liquid-Gas Solution. J.Non-Equilib.Thermodyn. 12 (1987)   (a)

SCHMELZER, J., SCHWEITZER, F.: On the Kinetic Description of Condensation in Binary Vapours, Ann.Physik 44 (1987) 283 (b)

SCHMELZER, J., SCHWEITZER, F.: Thermodynamics of Heterogeneous Systems:
Stability Analysis, Wiss.Z.WPU Rostock, NR 36 (1987)   (c)

SCHMELZER, J., SCHWEITZER, F.: Thermodynamics and Nucleation: II.
Adiabatic Nucleation in Finite Systems, Z.Phys.Chemie (Leipzig),
subm. for publication, 1987 (d)

SCHMELZER, J., ULBRICHT, H.: Thermodynamics of Finite Systems and the
Kinetics of First-Order Phase Transitions, J.Colloid Interface Sci.
    (1987)

SCHULTZ-PISZACHICH, W.: Nonlinear Models of Flow, Diffusion and Turbulence, Teubner-Texte zur Physik, Bd. 6, Leipzig, 1985

SCHWEITZER, F.: Thermodynamik und Keimbildung, Diplomarbeit, Wilhelm-
Pieck-Universität Rostock, 1983

SCHWEITZER, F.: Zur stochastischen Beschreibung der Keimbildung in
finiten Systemen, Diss. A, Wilhelm-Pieck-Universität Rostock, 1986
(a)

SCHWEITZER, F.: On the Kinetics of Nucleation: Stochastic Description
and Fokker-Planck-Equation, Wiss.Z.WPU Rostock, NR 35 (1986), H.4, 19
(b)

SCHWEITZER, F.: On the Droplet Formation in Dependence on Thermodynamic Constraints, Rostocker Phys.Manuskr. 9 (1986) 50 (c)

SCHWEITZER, F.: A Stochastic Approach to Nucleation in Finite Systems,
Rostocker Phys. Manuskr. 10 (1987)

SCHWEITZER, F., BUDDE, A., ULBRICHT, H.: On the Stochastic Droplet
Evolution in Finite Systems, in preparation

SCHWEITZER, F., SCHIMANSKY-GEIER, L.: Critical Parameters for Nucleation in Finite Systems, J.Colloid Interface Sci., (1987)

SCHWEITZER, F., SCHMELZER, J.: Thermodynamic Aspects of Nucleation,
Rostocker Phys. Manuskr. 10 (1987)

SCHWEITZER, F., TIETZE, H.: On the Nucleation Barrier in Finite Systems: An Investigation of the Thermodynamic Potential, Wiss.Z.WPU
Rostock, NR 36 (1987)

SCHWEITZER, F., ULBRICHT, H., SCHMELZER, J.: Thermodynamische Untersuchungen zu Phasenübergängen in finiten Systemen, Wiss.Z.WPU Rostock,
33 (1984), H.3, 45 (a)

SCHWEITZER, F., ULBRICHT, H., SCHMELZER, J.: Die Berechnung des er-
sten metastabilen Zustandes bei Kondensationsprozessen unter isochor-
isothermen Bedingungen, Wiss.Z.WPU Rostock 33 (1984), H.3, 54 (b)

SEKIMOTO, K.: Evolution of the Domain Structure During the Nucleation
and-Growth Process with Non-Conserved Order Parameter, Physica 135 A
(1986) 328

SHUGARD, W.J., REISS, H.: Transient Nucleation in $H_2O-H_2SO_4$-Mixtures:
A Stochastic Approach, J.Chem.Phys. 65 (1976) 2827

SIGGIA, E.D.: Late Stage of Spinodal Decomposition in Binary Mixtures,
Phys.Rev. A 20 (1979) 595

SKRIPOV, W.P., KOVERDA, V.P.: Spontane Kristallisation unterkühlter
Flüssigkeiten, Moskau 1984 (in Russian)

SPRINGER, G.S.: Homogeneous Nucleation, Adv.Heat Transfer 14 (1978) 281

SUBAREW, D.N.: Statistische Thermodynamik des Nichtgleichgewichts, Berlin, 1976

SUNQUIST, B.E., ORIANI, R.A.: Homogeneous Nucleation Theory in a Miscibility Gap System. A Critical Test of Nucleation Theory, J.Chem.Phys. 36 (1962) 2604

STANLEY , H.E.: Introduction to Phase Transitions and Critical Phenomena, Oxford, 1971

STEFAN, J.: Annalen der Physik 29 (1886) 655

TAMMANN, G.: Aggregatzustände, Leipzig, 1922

THOMPSON, S.M., GUBBINS, K.E., WALTON, J.P.R.B., CHANTRY, R.A.R., ROWLINSON, J.S.: A Molecular Dynamics Study of Liquid Drops, J.Chem. Phys. 81 (1984) 530

THOMSON, W.: Conduction of Electricity Through Gases, 2-nd Edition, Cambridge, 1906, p. 179ff., 3-rd Edition, Cambridge, 1929

TIMOSCHENKO, S.P., GOODIER, J.N.: Theorie der Elastizität, Moskau, 1975 (in Russian)

TODES, O.M.: Zh.Fiz.Khim. 20 (1946) 630

TODES, O.M., CHRUSCHTSCHEW, W.W.: Zh.Fiz.Khim. 21 (1947) 301

TOKUYAMA, M., KAWASAKI, K.: Statistical-Mechanical Theory of Coarsening of Spherical Droplets, Physica 123 A (1984) 386

TOKUYAMA, M., KAWASAKI, K., ENOMOTO, Y.: Kinetic Equations for Ostwald Ripening, Physica 134 A (1986) 323

TOLMAN, R.C.: Consideration of the Gibbs Theory of Surface Tension, J.Chem.Phys. 16 (1948) 758

TOLMAN, R.C.: The Superficial Density of Matter in a Liquid-Vapour Boundary, J.Chem.Phys. 17 (1949) 118

TOLMAN, R.C.: The Effect of Droplet Size on Surface Tension, J.Chem. Phys. 17 (1949) 333

TURSKI, L.A., LANGER, J.S.: Dynamics of a Diffusive Liquid-Vapor Interface, Phys.Rev. A 22 (1980) 2189

ULBRICHT, H., SCHWEITZER, F. (Eds.): Thermodynamics and Kinetics of First-Order Phase Transitions, Rostocker Phys.Manuskr. 10(1987)

ULBRICHT, H., SCHWEITZER, F., MAHNKE, R.: Nucleation Theory and Dynamics of First-Order Phase Transitions in Finite Systems. In: Self-organization by Nonlinear Irreversible Processes (Eds.: W. Ebeling, H. Ulbricht), Berlin (W.), 1986, p. 23

VAN DER WAALS, J.D.: Verhandel.Konink.Akad.Weten. Amsterdam (sect.1), vol.1 (1893) No.8; Z.Phys.Chem. 13 (1894) 657; van der Waals, Kohnstamm: Lehrbuch der Thermodynamik, Leipzig, Amsterdam, 1908; Rowlinson, J.S.: Translation of J.D. van der Waals' "The Thermodynamic Theory of Capillarity Under the Hypothesis of a Continuous Variation of Density", J.Stat.Phys. 20 (1979) 197

VAN KAMPEN, N.G.: Stochastic Processes in Physics, Lecture Notes, Utrecht, 1970

VENZL, G.: Ostwald Ripening of Precipitates. Numerical Solution for the Particle Size Distribution Function in Closed Systems, Ber.Bunsenges.Phys.Chemie 57 (1983) 318

VENZL, G.: Dynamics of First-Order Phase Transitions: Theory of Coarsening (Ostwald Ripening) for Open Systems, Phys.Rev. A 31 (1985) 3431

VOGELSBERGER, W.: Thermodynamische Untersuchungen zur Stabilität kleiner Flüssigkeitstropfen in Kontakt mit ihrem Dampf, Z.Phys.Chemie (Leipzig) 258 (1977) 763

VOGELSBERGER, W.: Thermodynamics of Finite Systems: A Possibility for Interpretation of Nucleation and Condensation Experiments, J.Colloid Interface Sci. 88 (1982) 17

VOGELSBERGER, W.: Thermodynamische und kinetische Untersuchungen zur Keimbildung und Kondensation - Erläutert am Phasenübergang Dampf - Flüssigkeit, Diss.B, Friedrich-Schiller-Universität Jena, 1983

VOGELSBERGER, W., MARX, G.: Zur Krümmungsabhängigkeit der Oberflächenspannung kleiner Tröpfchen, Z.Phys.Chemie (Leipzig) 257 (1976) 580

VOGELSBERGER, W., SONNEFELD, J., RUDAKOFF, G.: Theoretische und experimentelle Untersuchungen zur Kapillarkondensation - die Anwendung der Thermodynamik endlicher Systeme bei Berücksichtigung der Krümmungsabhängigkeit der Oberflächenspannung, Z.Phys.Chemie (Leipzig) 264 (1983) 625

VOGELSBERGER, W., SONNEFELD, J., RUDAKOFF, G.: Some General Considerations on a Curvature Dependent Surface Tension in the Capillarity of Liquids, Z.Phys.Chemie (Leipzig) 266 (1985) 225

VOLMER, M.: Kinetik der Phasenbildung, Dresden, 1939

VOLMER, M., WEBER, A.: Keimbildung in übersättigten Gebilden, Z.Phys. Chem. 119 (1926) 277

VOORHEES, P.W.: The Theory of Ostwald Ripening, J.Stat.Phys. 38 (1985) 231

VOORHEES, P.W., GLICKSMAN, M.E.: Solution to the Multi-Particle Diffusion Problem with Applications to Ostwald Ripening. I. Theory, II. Computer Simulations, Acta Met. 32 (1984) 2001, 2013

WAGNER, C.: Theorie der Alterung von Niederschlägen durch Umlösen (Ostwald-Reifung), Z.f.Elektrochemie 65 (1961) 581

WARD, C.A., LEVART, E.: Conditions for Stability of Bubble Nuclei in Solid Surfaces Contacting a Liquid-Gas Solution, J.Appl.Phys. 56 (1984) 491

WARD, C.A., TIKUISIS, P., VENTER, R.D.: Stability of Bubbles in a Closed Volume of Liquid-Gas Solutions, J.Appl.Phys. 53 (1982) 6076

WEGENER, P.P., MIRABEL, Ph.: Homogeneous Nucleation in Supersaturated Vapors, Naturwissenschaften 74 (1987) 111

WEGENER, P.P., WU, B.J.C.: Gasdynamics and Homogeneous Nucleation, In: Nucleation Phenomena (Ed.: A.C. Zettlemoyer), 1977, p. 325

WEISS, G.H.: Overview of Theoretical Models for Reaction Rates, J.Stat.Phys. 42 (1986) 3

WEST, B.J., LINDENBERG, K., SESHADRI, V.: Analytic Theory of Extrema: IV. Relaxation from metastable and unstable states, J.Chem.Phys. 72 (1980) 1151

WINGRAVE, J.A., SCHECHTER, R.S., WADE, W.H.: An Experimental Determination of the Curvature Dependence of Surface Tension from Fluid Flow Studies, In: The Modern Theory of Capillarity (Eds.: F.C. Goodrich, A.I. Rusanov), Berlin, 1981, p. 209

WONG, N.C., KNOBLER, C.M.: Light-Scattering Studies of Phase Separation in Isobutyric Acid and Watermixtures: Hydrodynamic Effects, Phys.Rev. A $\underline{24}$ (1981) 3205

WOODRUFF, D.P.: The Solid-Liquid Interface, Cambridge, Univ. Press, 1973

YANG, A.: Free Energy for the Heterogeneous System with Spherical Interfaces, J.Chem.Phys. $\underline{79}$ (1983) 6289

ZAHORANSKY, R.: Fortschritt-Berichte der VDI-Zeitschriften, Reihe 7, Nr. 70, Düsseldorf, 1982

ZELDOVITSCH, Yu.B.: Zur Theorie der Herausbildung einer neuen Phase Karitation, Zh.Exper.Teor.Fiz. $\underline{12}$ (1942) 525

ZETTLEMOYER, A.C. (Ed.): Nucleation, New York, 1969

ZETTLEMOYER, A.C.: Nucleation Phenomena, Adv.Colloid Interface Sci. $\underline{7}$ (1977)

ZUREK, W.H., SCHIEVE, W.C.: Vapor-Liquid Phase Transition in a Hard-Core Square-Well System: A Molecular Dynamics Study of Clustering and Nucleation. In: Phase Transitions, Vol. 2, 1981, p. 31

Diese Reihe wurde geschaffen, um eine schnellere Veröffentlichung
physikalischer Forschungsergebnisse und eine weitere Verbreitung
von physikalischen Spezialvorlesungen zu erreichen. TEUBNER-TEXTE
werden in deutsch oder englisch erscheinen. Um Aktualität der
Reihe zu erhalten, werden die TEUBNER-TEXTE im Manuskriptdruck
hergestellt, da so die geringeren drucktechnischen Ansprüche eine
raschere Herstellung ermöglichen. Autoren von TEUBNER-TEXTEN lie-
fern an den Verlag ein reproduktionsfähiges Manuskript. Nähere
Auskünfte darüber erhalten die Autoren vom Verlag.

This series has been initiated with a view to quicker publication
of the results of physical research-work and a widespread circula-
tion of special lectures on physics. TEUBNER-TEXTE will be published in German or English. In order to keep this series con-
stantly up to date and to assure a quick distribution, the copies
of these texts are produced by a photographic process (small-
offset printing) because its technical simplicity is ideally suited
to this type of publication. Authors supply the publishers with a
manuscript ready for reproduction in accordance with the latter's
instructions.

Cette série des textes a été créée pour obtenir une publication
plus rapide de résultats de recherches physiques et de conférences
sur des problèmes physiques spéciaux. Les TEUBNER-TEXTE seront
publiés en langues allemande ou anglaise. L'actualité des TEUBNER-
TEXTE est assurée par un procédé photographique (impression offset).
Les auteurs des TEUBNER-TEXTE sont priés de fournir à notre maison
d'édition un manuscrit prêt a être reproduit. Des renseignements
plus précis sur la forme du manuscrit leur sont donnéspar notre
maison.

Esta serie fué producida para lograr una publicación más rápida
de los resultados de la investigación en Fisica y una amplia
divulgación de las lecciones especiales en Fisica. Los TEUBNER-
TEXTE se editarán en idioma alemán o inglés. Para mantener la
actualidad de la serie, los TEUBNER-TEXTE se producen en imprenta
manuscriptal puesto que asi, las reducidas exigencias técnicas de
la elaboración posibilitan una producción rápida. Los autores de
los TEUBNER-TEXTE entregan a la Editorial manuscriptos aptos para
la reproducción. Los autores abtienen informaciones más detalladas
en la Editorial.

Эта серия была создана для обеспечения более быстрого публикования
результатов физических исследований и более широкого распростра-
нения физических лекций на специальные темы. Издания серии
ТОЙБНЕР-ТЕКСТЕ будут публиковаться на немецком или английском
языках. Для обеспечения актуальности серии, ее издания будут из-
готовляться фотомеханическом способом. Таким образом, более
скромные требования к полиграфическому оформлению обеспечат более
быстрое появление в свет. Авторы изданий серии ТОЙБНЕР-ТЕКСТЕ
будут предоставлять издательству рукописи, удовлетворящие требова-
ниям фотомеханического печатания. Более подробные сведения авторы
получат от издательста.

BSB B. G. Teubner Verlagsgesellschaft, Leipzig
DDR - 7010 Leipzig, Postfach 930